LECTURES ON PARTIAL DIFFERENTIAL EQUATIONS OF FIRST ORDER

ALBERTO DOU

UNIVERSITY OF NOTRE DAME PRESS

Lectures on Partial Differential Equations of First Order

LECTURES ON PARTIAL DIFFERENTIAL EQUATIONS OF FIRST ORDER

Alberto Dou

UNIVERSITY OF NOTRE DAME PRESS
NOTRE DAME LONDON

ISBN 0268-00468-4

Library of Congress Catalog Card Number: 74-186519

Printed in the United States by
NAPCO Graphic Arts, Inc., Milwaukee, Wisconsin

PREFACE

This book is based on lectures given by me at the University of Madrid and at the University of Notre Dame.

In writing the book I had three applications of the theory of partial differential equations of first order at the back of my mind: solving a large variety of problems of classical differential geometry; giving a systematic method for solving problems of the mechanical dynamics of a finite number of particles; and showing the connection of the theory with the study of the bicharacteristics of hyperbolic equations of higher order. Many instances of these applications will be found in the exercises.

It was my intention to publish a much larger book with an introduction to the theory of partial differential equations of second order, including motivation, existence and uniqueness and main properties of the solutions of equations with constant coefficients, and a discussion of the Cauchy Kovalevskaia theorem. But it will take still some time until this larger book is ready for print.

I wish to thank the Department of Mathematics of the University of Notre Dame for the encouragement that I received and the many facilities that were set at my disposal. My thanks go also to Mr. Michael A. Gauger for his help in writing down the English text, to Mrs. Bonnie Parsons for her patience and accuracy in typing the manuscript, and finally to the University Press for accepting the publication.

A. Dou

Notre Dame, Indiana

SYSTEMATIC INDEX

Chapter I

QUASI-LINEAR EQUATIONS

§ 1. Introduction

1. The Standard Problem.- We assume a knowledge of the theory of systems of ordinary differential equations, in particular of the existence and uniqueness of solutions of the initial value problem. The most important results of this theory are stated without proof at the end of this section. We recommend that the reader studying partial differential equations for the first time solve the exercises at the end of the section in order to familiarize himself with the basic concepts of the theory.

We will consider the partial differential equation of first order:

$$L(x,u,D)[u] \equiv L[u] \equiv$$

$$f_1(x,u)\frac{\partial u}{\partial x_1} + \dots + f_n(x,u)\frac{\partial u}{\partial x_n} - f(x,u) = 0, \tag{1,1}$$

where

$$x = (x_1,\dots,x_n),\ (x,u) \in D^{n+1} \subset E^{n+1}$$

$$f_1,\dots,f_n,\ f \subset C^1(D^{n+1}),\ D = \left(\frac{\partial}{\partial x_1},\dots,\frac{\partial}{\partial x_n}\right) \tag{1,2}$$

$$|f_1(x,u)| + \dots + |f_n(x,u)| > 0,\ (x,u) \subset D^{n+1},$$

and where u is the unknown to be determined. The equation is said to be quasi-linear since it is linear with respect to the first partial derivatives of u, although in general not linear with respect to u. The $(n + 1)$-dimensional Euclidean space is given together with the usual reference frame of Cartesian coordinates $\{0;\ x,u\}$ and D^{n+1} is a domain of this space E^{n+1}.

We will say that $u = \phi(x)$ is a solution or particular integral hypersurface (in explicit form) of equation (1) if the following two conditions are satisfied:

a) There exists a domain (open, connected, nonempty set) G of the Euclidean space E^n such that $\phi \in C^1(G)$ and

b) one has for $x \in G$,

$$(1,3) \qquad f_1(x, \phi(x)) \frac{\partial\phi}{\partial x_1} + \ldots + f_n(x,\phi(x)) \frac{\partial\phi}{\partial x_n} - f(x,\phi(x)) = 0.$$

This definition implies that if $x \in G$, then $(x,\phi(x)) \in D^{n+1}$.

We will now show in an intuitive and heuristic manner how one can expect that there are in fact integral surfaces of the equation (1).

2. Geometric Interpretation. - Let us consider the following vector field V associated with the points of D^{n+1}. Let $P = (x_1,\ldots,x_n,u) \in D^{n+1}$ and associate with it the vector $v(P) = (f_1,\ldots,f_n,f)$ where $f_1,\ldots,f_n,f$ are the components of the vector $v(P)$ with respect to the Cartesian coordinate system. We define characteristic curves of equation (1) to be the integral curves of the vector field V, that is, those curves which at each point P are tangent to $v(P)$. By virtue of the conditions (2), the characteristic curves fill the domain D^{n+1} in a regular fashion, that is, through each point $P \in D^{n+1}$ there passes a unique characteristic curve $\gamma(P)$.

For the sake of simplicity we assume $n = 2$ and we set $x = (x_1,x_2) = (x,y)$. Let Λ be a curve in $D^{n+1} = D^3$ with a continuous tangent that is transverse to the characteristic curve, that is at no point P of Λ is the direction of the tangent to Λ through P equal to the direction of $v(P)$. Now construct for each point $P \in \Lambda$ the characteristic curve $\gamma(P)$. Thus we have generated a surface Σ which is an integral surface of (1) in a neighborhood of Λ. That is to say, there exists a domain $G \subset E^2 = \{0;\ x,y\}$ which contains the projection Λ^* of Λ into the plane $\{0;\ x,y\}$ and in which the integral surface Σ is defined.

We shall indicate why. As a consequence of the regularity conditions imposed on the curve Λ and on the operator L, the surface Σ has a tangent plane at each of its points. Now if $M \in \Sigma$, the tangent plane to Σ through M contains the tangent line through M to the characteristic curve $\gamma(M)$. Therefore the normal through M to Σ is orthogonal to the vector $v(M)$; hence the scalar product of the two is zero for all points of the surface Σ. But assuming that $u = \phi(x,y)$ is the equation of Σ, the direction parameters of the normal to Σ are $(\frac{\partial u}{\partial x}, \frac{\partial u}{\partial y}, -1)$ and therefore, the scalar product being zero, one obtains:

$$\left\langle \left(\frac{\partial u}{\partial x}, \frac{\partial u}{\partial y}, -1 \right), (f_1(x,y,u), f_2(x,y,u), f(x,y,u)) \right\rangle =$$

$$f_1(x,y,u) \frac{\partial u}{\partial x} + f_2(x,y,u) \frac{\partial u}{\partial y} - f(x,y,u) = 0. \tag{1,4}$$

That is, the two conditions of (3) are fulfilled, q.e.d.

It is obvious that the construction we have given applies equally to the case $n > 2$. One replaces Λ with an $(n-1)$-dimensional manifold whose tangent hyperplane is continuous and transverse to the characteristic curves. Other secondary assumptions will be given precisely later on. Constructing a characteristic curve through each point of this transverse manifold, we generate an n-dimensional integral hypersurface of the equation (1). The justification is essentially the same as before.

In the case $n = 2$, the integral surface is generated by a 1-parameter system of characteristic curves, and in the general case, in the same manner, the integral hypersurface is generated by an $(n-1)$-parameter system of characteristic curves. When $n > 2$, in addition to the integral hypersurfaces or n-dimensional integral manifolds, there will be other m-dimensional integral manifolds where m is any natural number satisfying $2 \leqq m \leqq n - 1$. The definition of an m-dimensional integral manifold is similar to that of an integral hypersurface, except that now G is m-dimensional in such a way that $x_1,..,x_n$ depend only on m-parameters or independent variables. One can construct an m-dimensional integral manifold, $2 \leqq m \leqq n$, proceeding as before, that is, beginning with an $(m-1)$-dimensional manifold transverse to the characteristic curves and constructing through each point of this manifold the characteristic curve. Thus one generates an $(m-1)$-parameter system of characterisitic curves which constitutes the m-dimensional integral manifold.

Let Σ,

$$\Sigma: \quad u = \phi(x), \ x \in G ,$$

be an integral hypersurface of the equation (1) and $M \in \Sigma$. From the uniqueness of the characteristic curve $\gamma(M)$ through M and considering that according to (4) the vector $v(M)$ must lie in the tangent hyperplane through M to Σ, it seems highly intuitive and likely (and we shall see later it is so) that the curve $\gamma(M)$ must lie in Σ , provided the projection γ^* (M) of $\gamma(M)$ lies in G. Hence, one may deduce that if two integral surfaces intersect at a point, they also intersect everywhere along the characteristic curve through this point. And if they are tangent at a point, they are also tangent everywhere along the characteristic curve through this point.

3. Characteristic Curves.- For characteristic curves, $(f_1(x,u),\ldots, f_n(x,u), f(x,u))$ as well as $(dx_1,\ldots, dx_n, du)$ are direction parameters for the tangent. Therefore, the ordinary differential system of characteristic curves is

$$(1,5a)\quad \frac{dx_1}{f_1(x,u)} = \ldots = \frac{dx_n}{f_n(x,u)} = \frac{du}{f(x,u)}\ , \quad (x,u) \in D^{n+1}\ .$$

Sometimes it becomes convenient to introduce an auxilliary parameter s, and the system then becomes

$$(1,5b)\quad \frac{d\,x_1}{f_1(x,u)} = \ldots = \frac{d\,x_n}{f_n(x,u)} = \frac{du}{f(x,u)} = ds.$$

The characteristic curves are the trajectories of the system (5a), or equivalently, the paths of the autonomous system (5b). The system is of order n, and therefore there exists an n-parameter system of characteristic curves. In the system (5b), the constant of integration s_o may be associated with the variable s giving $s - s_o$, since the system is autonomous, that is the system does not depend on s.

We define base characteristic curves as the projection curves of the characteristic curves on the hyperplane $\{0; x\}$. In general, the projection of characteristic curves on the base characteristic curves is bijective and therefore there exists an n-parameter system of base characteristic curves. But, if the equation (1) is semilinear, that is if the functions $f_1,..,f_n$ depend only on the independent variables $x_1,...,x_n$ and not on u, then the set of base characteristic curves depends only on n - 1 parameters and is defined by the first n - 1 equations of (5a).

For an integral surface of the quasi-linear equation (1), for instance for $u = \phi(x)$, $x \in G$, the set of base characteristic curves, projections of the characteristic curves which generate the integral surface, is given by the first n - 1 equations of system (5a) when we substitute $\phi(x)$ for u. It should be remarked that no characteristic curve may have a tangent parallel to the u-axis due to the conditions (2).

Finally, without proof, we give a classification of the characteristic curves in accordance with their geometric properties. Let σ^+ be a semipath of the system (5b), that is, an oriented half trajectory of system (5a) in the domain D^{n+1}. We set

$$\sigma^+ : \begin{cases} x = \xi(s) \\ u = \zeta(s), \; s \geqq s_o , \end{cases} \qquad \begin{cases} \xi(s_o) = x^o \\ \zeta(s_o) = u_o , \end{cases}$$

in such a way that σ^+ is the positive semipath through the point $(x^o, u_o) \in D^{n+1}$. What is said for the positive semipath σ^+ holds likewise for the negative semipath σ^-. We consider the semipath σ^+ both as a solution of the differential system (5) and as a point set in the domain D^{n+1}, that is

$$\sigma^+ = \{(x,u) | \; x = \xi(s), \; u = \zeta(s), \; s \geqq s_o\} .$$

To simplify the expression we define the external boundary of σ^+, relative to the Euclidean space $\{O; x,u\}$, as the set of points which belong to the boundary of σ^+ but not to σ^+. Then each of the semipaths is a member of one and only one of the following four categories:

a) The semipath σ^+ has a non-empty external boundary which is contained in the boundary of D^{n+1}.

b) The semipath σ^+ has an empty external boundary and is not bounded. Of course this case is not possible if D^{n+1} is bounded.

c) The semipath σ^+ has empty external boundary and is bounded. In this case the semi-path σ^+ and the entire path σ are closed and the functions ξ and ζ are periodic with the same period.

d) The semipath σ^+ has a non-empty external boundary which is not contained in the boundary of D^{n+1}. In this case σ^+ is bounded, but not closed, has infinite length and an external boundary K which is a closed and connected point set contained in the closure of D^{n+1}. If $M \in K \cap D^{n+1}$, then the entire path through M is contained in K and is a limit path.

Exercises

1. The equation $u = f(x^2 + y^2)$, where f is an arbitrary function, represents the set of surfaces of revolution whose axis coincides with the u coordinate axis and one of whose meridians is $u = g(x) = f(x^2)$.

Taking partial derivatives with respect to x and y, one can eliminate f and obtain the partial differential equation

$$y \frac{\partial u}{\partial x} - x \frac{\partial u}{\partial y} = 0.$$

So this equation also represents the same set of surfaces. Find the characteristic curves of this equation and verify that the integral surface through any curve is in fact a surface of revolution whose axis is the u coordinate axis.

Compare the elimination of the arbitrary function f with the elimination of an arbitrary constant in the case of a 1-parameter system of curves, i.e. when there is only one independent variable.

2. The same exercise as (1), but starting from the equation $f(x^2 - y^2) = u$.

3. The same exercise as (1), but beginning with the equation $u = f(a(x,y))$ where $a(x,y)$ is a determined function.

4. Find the characteristic curves of the equation

$$m \frac{\partial u}{\partial x} + n \frac{\partial u}{\partial y} - p = 0,$$

where m, n and p are real constants.

Verify that the characteristic curves are precisely the family of straight lines parallel to the line

$$\frac{x}{m} = \frac{y}{n} = \frac{u}{p} \;.$$

Therefore the general solution of the equation is the set of cylinders whose generators are parallel to the given direction, i.e.

$$W(nx - my \;, \; py - nu) = 0.$$

Find the particular solution of the equation which passes through the curve

$$y = 0, \ u = f(x) \ .$$

5. Verify that the elimination of the arbitrary function w of the equation

$$F(x,y,u, w[a(x,y,u)]) = 0 \quad ,$$

leads to a quasi-linear partial differential equation of first order. The functions F and a are assumed to be known.

6. Generalize the Exercise (5) to the case of n independent variables $x_1, \ldots, x_n$. Remark that w must now be a function of n-1 arguments.

7. Given the equation

$$u = f\left(\frac{xy}{u} \right) \quad ,$$

eliminate the arbitrary function f.

8. Find the partial differential equation of all surfaces of revolution with axis the line

$$\frac{x}{a} = \frac{y}{b} = \frac{u}{c} \quad .$$

9. Find the partial differential equation of all conical surfaces with vertex the point (a,b,c).

10. Let

$$V = (f(x,y,z), \ g(x,y,z), \ h(x,y,z))$$

be a vector field in a domain of the Euclidean space $\{0; x,y,z\}$. Let $z = w(x,y)$ be a surface whose tangent plane thru the point $(x,y,w(x,y,))$ contains the vector whose origin lies at this point and whose components are $f(x,y,w(x,y))$, $g(x,y,w(x,y))$, $h(x,y,w(x,y))$; otherwise w is arbitrary.

Find the partial differential equation of first order in terms of the functions f, g, h that the function $z = w(x,y)$ satisfies and justify the answer.

11. Let $x = \xi(s; s_o, x^o, u_o)$, $u = \zeta(s; s_o, x^o, u_o)$ be the general solution of the differential system (5b) of the characteristic curves. Prove that

$$\xi(s\ ;\ s_o,\ x^o,\ u_o) = \xi(s - s_o\ ;\ 0,\ x^o,\ u_o)$$

$$\zeta(s\ ;\ s_o,\ x^o,\ u_o) = \zeta(s - s_o\ ;\ 0,\ x^o,\ u_o) \quad .$$

12. Find the partial differential equation which results from eliminating the arbitrary function w from the equation

$$u = w(a(x,y)),$$

where $a(x,y)$ is assumed to be known.

Verify that one obtains

$$\frac{\partial(w,a)}{\partial(x,y)} = 0 \quad .$$

Hence the equation $u = w(a(x;y))$ is equivalent to the partial differential equation.

Considering the result of Exercise (5), generalize the present result (i.e. the equivalence) to the case in which a depends on x, y, u. Moreover, consider the particular case in which a is given by

$$a(x,\ y,\ u) = x\cdot g(u) - y\cdot f(u).$$

Verify that in this particular case one obtains the equation

$$x\cdot g(u) - y\cdot f(u) - w(u) = 0$$

as the general solution of

$$f(u)\ \frac{\partial u}{\partial x} + g(u)\ \frac{\partial u}{\partial y} = 0.$$

13. Assume that $u(x_1,\ \dots\ ,\ x_n)$ is a homogeneous function of degree k, k real, that is

$$u(tx_1,\dots,tx_n) = t^k\cdot u(x_1,\dots,x_n),\ t \in \mathbb{R}.$$

Prove the Euler formula concerning homogeneous functions, i.e.

$$x_1\ \frac{\partial u}{\partial x_1} + \dots + x_n\ \frac{\partial u}{\partial x_n} = k\ u.$$

[To this end, one can take partial derivatives with respect to t and then set t = 1. Also, see Exercise 4. at the end of § 2].

§ 2. Integral Surfaces and First Integrals

1. The Jacobi Transformation. - We will consider the quasi-linear equation with two independent variables

$$(2,1)\quad N[u] \equiv f(x,y,u)\frac{\partial u}{\partial x} + g(x,y,u)\frac{\partial u}{\partial y} - h(x,y,u) = 0$$

satisfying the conditions (1,2) for $n = 2$.

Let $u = \phi(x,y)$ be a solution of (1) in a neighborhood G_0 of (x_0,y_0). It is understood that G_0 is an open, simply connected set containing (x_0,y_0), and as small as desired. Theorems concerning domains which are arbitrarily small neighborhoods of a point are called local theorems, that is, the domain in which the theorem holds, which of course must be shown to exist, is allowed to be as small a neighborhood of (x_0,y_0) as may be needed. Such will be the theorems of this section in order that the statements may be simpler. Ordinarily it is not possible to generalize such theorems to global results.

In this § 2, we propose to see the relation between the equation (1) and its characteristic system

$$(2,2)\qquad \frac{dx}{f(x,y,u)} = \frac{dy}{g(x,y,u)} = \frac{du}{h(x,y,u)}\ .$$

This system of equations is symmetric with respect to the three variables x,y,u which is not the case in equation (1). It is not difficult to establish an equation closely related to (1) which is symmetric with respect to x,y,u. This is precisely the purpose of the Jacobi transformation.

To this end, instead of particular integrals like $u = \phi(x,y)$, we must consider 1-parameter systems of integral surfaces.

Now, let $\phi(x_o,y_o) = u_o$ and D_o be a neighborhood in D^3 of the point (x_o,y_o,u_o). Let

$$(2,3)\quad w(x,y,u) = c\ ,\qquad \frac{\partial w(x_o,y_o,u_o)}{\partial u} \neq 0\ ,\ w \in C^1(D_o),$$

be a 1-parameter system of integral surfaces in a neighborhood of (x_o,y_o) depending on the parameter c, that is, for each value of c, the funtion $u = \psi(x,y,c)$ implicitly defined by (3) is an integral surface in G_o. The parameter c assumes real values in an open neighborhood of $c_o = w(x_o,y_o,u_o)$; moreover, we may assume, without restriction, that $\psi(x,y,c_o) = \phi(x,y)$.

Assertion 2.1. *If* $w(x,y,u)$, where

$$\frac{\partial w(x_o,y_o,u_o)}{\partial u} \neq 0\ ,\quad w \in C^1(D_o)\ ,$$

is such that $w(x,y,u) = c$ *defines implicitly a 1-parameter system of surfaces in* G_o *of* (1), *then* $w = w(x,y,u)$ *is a solution in* D_o *of the equation*

$$(2,4)\quad J[w] \equiv f(x,y,u)\,\frac{\partial w}{\partial x} + g(x,y,u)\,\frac{\partial w}{\partial y} + h(x,y,u)\,\frac{\partial w}{\partial u} = 0.$$

Conversely, if $w = w(x,y,u)$ *is a solution of* (4) *in* D_o *and* $\frac{\partial w(x_o,y_o,u_o)}{\partial u} \neq 0$, *then* $w(x,y,u) = c$ *implicitly defines a 1-parameter system of integral surfaces in* G_o *of* (1).

The change of variables u into w by $w = w(x,y,u)$, that is the transformation of (1) into (4), is called the Jacobi transformation. Equation (4) is linear and homogeneous in two ways; it has no second member and all the terms contain a partial derivative of first order. Moreover, notice that (4) is symmetric with respect to the three independent variables x,y,u.

Proof of the assertion. From (3), we conclude that for any value of c

$$\frac{\partial w}{\partial x} + \frac{\partial w}{\partial u} \cdot \frac{\partial u}{\partial x} = 0 , \quad \frac{\partial w}{\partial y} + \frac{\partial w}{\partial u} \cdot \frac{\partial u}{\partial y} = 0. \tag{2,5}$$

If by (5) we eliminate $\frac{\partial u}{\partial x}$, $\frac{\partial u}{\partial y}$ in the first member of (1), we obtain precisely $J[w]$ of the first member of (4). Now considering that u, implicitly defined by $w(x,y,u) = c$, is a solution of (1) in a neighborhood G_o depending on c, we conclude that w satisfies the equation $J[w] = 0$ in D_o, q.e.d.

Conversely. Assume $w = w(x,y,u)$ is a solution of (4) in D_o, and that

$$\frac{\partial w(x_o,y_o,u_o)}{\partial u} \neq 0 .$$

Then $w(x,y,u) = c$ implicitly defines a 1-parameter system of functions satisfying (5) for all c is a neighborhood of c_o. Then by (5), we may pass from equation (4) to equation (1) and conclude that the functions $u = \psi(x,y,c)$, $(x,y) \in G_o$, implicitly defined by $w(x,y,u) = c$ are solutions of (1), q.e.d.

The Jacobi transformation allows us to pass from a quasi-linear equation (1) to a linear and particularly simpler one such as (4) at the cost of adding one independent variable. In the same way, the Jacobi transformation induces a bijective correspondence between 1-parameter systems of local solutions

$$w(x,y,u) = c \quad , \quad (x,y) \in G_o ,$$

of (1) and local solutions

$$w = w(x,y,u) \quad , \quad (x,y,u) \in D_o ,$$

of (4).

The generalization of the present assertion for equation (1) with two independent variables to equation (1,1) with n independent variables is obvious.

2. First Integrals of Characteristic Systems.- Having seen the relation between (1) and (4), we now study the characteristic system (2).

Consider a general system of n ordinary differential equations in symmetric form:

$$\frac{dx_1}{f_1(x)} = \frac{dx_2}{f_2(x)} = \dots = \frac{dx_{n+1}}{f_{n+1}(x)} ,$$

(2,6a) $\qquad x = (x_1, x_2, \dots , x_{n+1}) ,$

(2,6b) $\qquad f_1, \dots , f_{n+1} \in C^1(D^{n+1})$

$$| f_1(x) | + \dots + |f_{n+1}(x)| > 0 , \ x \in D^{n+1} ,$$

where D^{n+1} is a domain of E^{n+1}. The last condition insures that for any point $x^o \in D^{n+1}$ the system (6) may be put in a normal and regular form. That is, if $f_j(x^o) \neq 0$, taking x_j as an independent variable, one can put (6) in a normal form satisfying sufficient conditions for the existence and uniqueness of the trajectories in D^{n+1} in a neighborhood of x_j^o

on the x_j-axis.

One says that the function

$$w(x) \ , \ x \in \hat{D}^{n+1} \subset D^{n+1} \quad ,$$

where $\hat{D}^{n+1}$ is a domain of E^{n+1}, is a first integral of system (6), provided that $w \in C^1(\hat{D}^{n+1})$, $w(x)$ remains constant as x varies along any trajectory in $\hat{D}^{n+1}$, and that all its derivatives $\frac{\partial w}{\partial x_i}$ do not simultaneously vanish. The last condition, besides excluding trivial constant functions, is necessary to insure the second condition of (3) which is needed. Saying that $w(x)$ is constant along any trajectory, we imply that this constant may depend on the trajectory.

The differential system of order n in the symmetric form (6) is the base characteristic system of the following partial differential equation of first order with (n+1) independent variables in Jacobi form,

$$J\,[\,w\,] \equiv f_1(x)\,\frac{\partial w}{\partial x_1} + \dots + f_{n+1}(x)\,\frac{\partial w}{\partial x_{n+1}} = 0\ , \qquad (x,w) \in D^{n+1} \times E^1\ ; \tag{2,7}$$

the conditions (6b) are precisely the conditions (1,2) corresponding to equation (7). Observe that in accordance with the definition given in § 1, article 3, the characteristic system of (7) is

$$\frac{dx_1}{f_1(x)} = \dots = \frac{dx_{n+1}}{f_{n+1}(x)} = \frac{dw}{0} \quad .$$

The relation between the equation in Jacobi form (7) and its base characteristic system (6) is given by the following theorems.

Theorem 2.1. Let $w(x)$ be a first integral of (6) for $x \in \hat{D}^{n+1} \subset D^{n+1}$. Then $w = w(x)$ is an integral surface of (7) for $x \in \hat{D}^{n+1}$.

Conversely: If $w = w(x)$ is an integral surface of (7) for $x \in \hat{D}^{n+1} \subset D^{n+1}$, then $w(x)$ is a first integral of (6) in $\hat{D}^{n+1}$.

Proof of the first part. Let $x^{o} \in \hat{D}^{n+1}$ and let γ_{o} be the trajectory of (6) passing through x^{o}. Since $w(x)$ is a first integral in $\hat{D}^{n+1}$, it will remain constant as x moves along γ_{o}, that is to say

$$(2,8) \qquad dw(x) \equiv \frac{\partial w}{\partial x_1} dx_1 + \dots + \frac{\partial w}{\partial x_{n+1}} dx_{n+1} = 0.$$

Now, precisely as x moves on a trajectory, equation (6) holds, and hence (8) implies, for $x = x^{o}$

$$(2,9) \qquad f_1(x^{o}) \frac{\partial w(x^{o})}{\partial x_1} + \dots + f_{n+1}(x^{o}) \frac{\partial w(x^{o})}{\partial x_{n+1}} = 0 .$$

But since x^{o} is arbitrary in $\hat{D}^{n+1}$, it follows from (9) that $w = w(x)$ is an integral surface of (7), q.e.d.

Proof of the converse. This result is trivial, since as x moves on a trajectory of (6) one has

$$\frac{dx_1}{f_1(x)} = \dots = \frac{dx_{n+1}}{f_{n+1}(x)} = \frac{\frac{\partial w}{\partial x_1} dx_1 + \dots + \frac{\partial w}{\partial x_{n+1}} dx_{n+1}}{f_1(x)\frac{\partial w}{\partial x_1} + \dots + f_{n+1}(x) \frac{\partial w}{\partial x_{n+1}}} ;$$

but, since $w = w(x)$ is an integral surface of (7), the last denominator is zero, and therefore also the numerator; that is $dw(x) = 0$ when x moves on a trajectory. Hence $w(x)$ is constant and therefore a first integral of (6), q.e.d.

Theorem 2.2. *If*

$$w = w_1(x), \; \dots \; , w = w_k(x) \; , \; 1 \leqq k \leqq n,$$

are integral surfaces of (7) *in a neighborhood* D_o^{n+1} *of* x^o, *and if* $Z(w_1, \dots , w_k)$ *is such that* $Z \in C^1(B_o^k)$, *where* B_o^k *is a neighborhood of* $(w_1(x^o), \dots , w_k(x^o)) \equiv (w_{1o}, \dots , w_{ko})$, then $w = Z(w_1(x), \dots , w_k(x))$ *is an integral surface of* (7) *in a neighborhood of* x^o.

Let $k = n$ *and* $w = w_1(x), \dots , w = w_n(x)$ *be integral surfaces of* (7) *in* D_o^{n+1} *and assume that they are functionally independent, that is, the Jacobian matrix*

$$(2,10) \qquad J(x^o) \equiv \left(\frac{\partial(w_1, \dots , w_n)}{\partial(x_1, \dots , x_{n+1})} \right)_{x = x^o}$$

has rank n. *If* $w = w_{n+1}(x)$ *is an integral surface of* (7) *in a neighborhood of* x^o, *then there exists a function* $Z \in C^1(B_o^n)$ *such that*

$$Z(w_1(x), \dots , w_n(x)) = w_{n+1}(x) \; , \; x \in D_o^{n+1} .$$

Proof of the first part. Substituting $Z(w_1(x)), \dots , w_k(x))$ in (7), one obtains

$$J\,[Z(x)] = \frac{\partial Z}{\partial w_1} J\,[w_1] + \dots + \frac{\partial Z}{\partial w_k} J\,[w_k] \; , \; x \in D_o^{n+1} ,$$

but since $w = w_1, \dots , w = w_k$ are integral surfaces one has $J\,[w_1] = \dots = J\,[w_k] = 0$, and hence $J\,[Z(x)] = 0$, $x \in D_o^{n+1}$, that is to say, $Z(x)$ is an integral surface, q.e.d.

Proof of the second part. Since $w = w_i(x)$, $z = 1, \dots ,$ $n+1$, are integral surfaces of (7), one has

$$J\,[w_i] = \frac{\partial w_i}{\partial x_1} f_1(x) + \dots + \frac{\partial w_i}{\partial x_{n+1}} f_{n+1}(x) = 0,$$

(2,11) $\qquad i = 1, \dots, n+1 , x \in D_o^{n+1}$.

Now for any $x \in D_o^{n+1}$, the equations (11) may be considered as a linear algebraic system of n+1 equations in the n+1 unknowns $f_1(x), \dots, f_{n+1}(x)$, which in fact has a non-trivial solution because of (6b). Therefore the determinant of the coefficients must be zero for all $x \in D_o^{n+1}$. But this determinant is precisely the Jacobian of $w_1, \dots, w_{n+1}$, hence one has

$$\frac{\partial(w_1, \dots, w_{n+1})}{\partial(x_1, \dots, x_{n+1})} = 0 , \quad x \in D_o^{n+1} \quad .$$

It implies that there is a functional relation between the functions $w_1, \dots, w_{n+1}$. But $w_1, \dots, w_n$ are functionally independent in D_o^{n+1}, hence there exists Z such that $w_{n+1}(x) = Z(w_1(x), \dots, w_n(x))$, q.e.d.

By virtue of Theorem 2.1, it is clear that the properties we have just proven for the integral surfaces $w_k = w_k(x)$ translate easily into properties of the first integrals $w_k(x)$. In particular, let us assume that the minor in the matrix $J(x^o)$ formed by the first n columns is nonsingular, and set $(x_1, \dots, x_n) = \bar{x}$, $x_{n+1} = s$, $(x_1^o, \dots, x_n^o) = \bar{x}^o$, $x_{n+1}^o = s_o$.

Then we can solve the system

$$w_i(x) = w_i(x^o), \; i = 1, \dots, n , \; x \in D_o^{n+1} ,$$

for the unknown functions $x_1(s), \dots, x_n(s)$, obtaining

$$\bar{x} = \phi(s \; ; \; s_o, \bar{x}_o) \quad ,$$

which is precisely the general integral of (6) in a neighborhood of the point $(s_o, \bar{x}^o)$.

Let $W \in C^1(B_o^n)$, otherwise arbitrary, B_o^n being the neighborhood of $(w_{10}, \ldots, w_{n0})$ defined in theorem 2.2, and let $w_1(x), \ldots, w_n(x)$ be functionally independent. Then the expression $W = W(w_1(x),\ldots,w_n(x))$ is called the <u>general integral</u> of (7) in a neighborhood of x^o. The sense of the term "general" is of course the one given in the second part of Theorem 2.2. One might also call it a general integral because according to Theorem 3.2 it permits us to solve the Cauchy problem for any initial curve through the point (x^o,w_o). Observe that these integral surfaces of (7) with n+1 independent variables or first integrals of the system (6) of order n, contain an arbitrary function W which depends on n independent arguments.

Finally, we refer to the exercises to see how to find first integrals of a system such as (6), and to better understand their importance in reducing the order of the system in pursuit of an effective solution. In the same way we refer to the exercises to see how the knowledge of k functionally independent surfaces of (7) allows through a change of independent variables, a reduction of (7) to an equation also in Jacobi form, but with only n+1-k independent variables.

Exercises

1. Verify that the function

$$w(x,y,u) = xy - u^2$$

is a first integral of the system

$$\frac{dx}{x} = \frac{dy}{2xu-y+2u^3} = \frac{du}{x^2 + u^2} .$$

2. Let $a_1(x,u), \ldots , a_n(x,u), a(x,u)$ be a family of factors of the differential system (1,5a), that is

$$a_1(x,u) \cdot dx_1 + \ldots + a_n(x,u) \cdot dx_n + a(x,u)\, du \equiv dw(x,u),$$

(hence the first member is an exact differential) such that simultaneously

$$a_1 f_1 + \ldots + a_n f_n + af \equiv 0 .$$

   a) Prove that $w(x,u)$ is a first integral of the system.

   b) Prove that if $f_i = f_i(x_j)$ and $f_j = f_j(x_i)$, then there exists a first integral $w(x_i,x_j)$ depending only on x_i, x_j.

3. Find the general integral of the equation

$$(x^2 + xy) \frac{\partial u}{\partial x} - (xy + y^2) \frac{\partial u}{\partial y} = (y - x)(2x + 2y + z).$$

4. Let $u(x_1, \ldots , x_n)$ be a homogeneous function of degree k, k real, that is

$$u(tx_1, \ldots , tx_n) = t^k u(x_1, \ldots , x_n).$$

Prove that u satisfies the Euler formula

$$x_1 \frac{\partial u}{\partial x_1} + \ldots + x_n \frac{\partial u}{\partial x_n} = ku ,$$

and that conversely, any solution of this equation is a homogeneous function of degree k.

5. Prove that in order for a surface $u = \phi(x,y)$ to be orthogonal to a given family of surfaces $F(x,y,u) = h$ it is necessary and sufficient that

$$F_x \frac{\partial\varphi}{\partial x} + F_y \frac{\partial\varphi}{\partial y} = F_u .$$

Find the general equation of all surfaces orthogonal to the parabaloids $x^2 + y^2 = 2\,hu$.

6. Let $F(x,y,u) = k$ be a 1-parameter system of surfaces.

a) Prove that the curve $x = x(t)$, $y = y(t)$, $u = u(t)$ is orthogonal to the system of surfaces if and only if

$$\frac{dx(t)}{F_x(x(t),y(t),u(t))} = \frac{dy}{F_y} = \frac{du}{F_u} .$$

b) Find the field curves $x(t)$, $y(t)$, $u(t)$ of the conservative vector field whose potential function is

$$F(x,y,u) = x^2 + y^2 + u^3.$$

7. Let $x = (x_1, \ldots , x_n)$ and let $\phi(x) = (\phi_1, \ldots , \phi_{n-1})$ be a functionally independent set of n-1 functions of x. Prove that a partial differential equation has $\phi_1(x), \ldots , \phi_{n-1}(x)$ as solutions if and only if the equation has the form

$$g(x) \cdot \frac{\partial(u,\phi)}{\partial(x)} = 0 ,$$

where $g(x)$ is arbitrary.

8. Suppose we are given k functionally independent integral surfaces $w_1 = w_1(x), \ldots , w_k = w_k(x)$ of equation (7). Prove that if one preserves the variables $w, x_1, \ldots , x_{n+1-k}$ and makes the change

$$w_1(x) = y_1, \ldots , w_k(x) = y_k ,$$

of the independent variables $x_{n-k+2}, \ldots, x_{n+1}$ into the variables $y_1, \ldots, y_k$, equation (7) becomes

$$g_1(x^*,y)\frac{\partial w}{\partial x_1} + \ldots + g_{n+1-k}(x^*,y)\frac{\partial w}{\partial x_{n+1-k}} = 0 ,$$

where $x^* = (x_1, \ldots, x_{n-k+1})$, $y = (y_1, \ldots, y_k)$.

[See that the transformed equation must be of the same Jacobi form and has $w = y_i$, $i = i, \ldots, k$ as solutions.]

9. Assume we are given k functionally independent first integrals $w_1(x), \ldots, w_k(x)$ of the differential system (6). Reduce the system by means of a change of independent variables to a system of order (n-k). [See preceeding exercise.]

10. Assume that $w(x,y,u)$ is a first integral of the system

$$dx = \frac{dy}{u} = \frac{du}{f(x,y)} ,$$

where $f(x,y)$ is also given. Let $u = \sigma(x,y,C)$ be the function implicitly defined by $w(x,y,u) = C$, C a constant.

Prove that the partial derivative of u with respect to C is an integrating factor of the equation

$$\sigma(x,y,C)\,dx - dy = 0 ,$$

which corresponds to the first equation of the system.

Apply this result to the solution of the second order equation

$$\frac{d^2y}{dx^2} = f(x,y) .$$

11. Given the system

$$dx = \frac{dy}{zu} = \frac{dz}{-uy} = \frac{du}{-k^2yz} , \quad 0 < k < 1 ,$$

find the integral curve thru the point

$$(x_0, y_0, z_0, u_0) = (0, b, c, k^2 c).$$

[Verify that one obtains strictly elementary functions precisely because of the special initial condition.]

§ 3. The Cauchy Problem

1. An Uniqueness Theorem.- In the geometric interpretation given in the introduction, we intuitively showed that there exist integral surfaces, and we have pointed out some of their properties. We now propose to prove that interpretation is correct. Consider the quasi-linear equation,

$$L(x,u,D)[u] \equiv L[u] \equiv$$

$$f_1(x,u)\frac{\partial u}{\partial x_1} + \dots + f_n(x,u)\frac{\partial u}{\partial x_n} - f(x,u) = 0, \tag{3,1}$$

where

$$x = (x_1, \dots, x_n),\ (x,u) \in D^{n+1} \subset R^{n+1}$$

$$f_1, \dots, f_n, f \in C^1(D^{n+1}) \tag{3,2}$$

$$|f_1(x,u)| + \dots + |f_n(x,u)| > 0,\ (x,u) \in D^{n+1},$$

and its characteristic system

$$\frac{dx_1}{f_1(x,u)} = \dots = \frac{dx_n}{f_n(x,u)} = \frac{du}{f(x,u)} = ds. \tag{3,3}$$

Before we approach the Cauchy theorem, we give a preliminary result which will be useful in establishing the unicity of the solution.

Theorem 3.1. Let Σ, defined by $u = \phi(x)$, be an integral surface of (1) for $x \in G$. Let $\gamma \subset D^{n+1}$ be a characteristic curve through (x^o, u_o), $u_o = \phi(x^o)$, and such that $\gamma^* \subset G$ where γ^* is the projection of γ on $\{0; x\}$. Then $\gamma \subset \Sigma$.

Proof. Since γ is a characteristic curve through (x^o, u_o), its equations are the unique solutions of system (3) which for $s = s_o$ pass through (x^o, u_o):

$$x = \xi(s;\ s_o, x^o, u_o) \equiv \xi(s), \xi(s_o) = x^o$$

$$(3,4) \qquad u = \zeta(s;\ s_o, x^o, u_o) \equiv \zeta(s), \zeta(s_o) = u_o,$$

$$-\ \eta_1 < s - s_o < \eta_2;\ \eta_1, \eta_2 > 0.$$

Consider the curve δ defined by:

$$x = \hat{\xi}(s)$$

$$u = \phi(\hat{\xi}(s)) \equiv \psi(s),\ \psi(s_o) = \varphi(x^o) = u_o,$$

where $x = \hat{\xi}(s)$ is defined as the unique solution through x^o of the base characteristic system when one substitutes $\phi(x)$ for u in the functions $f_1(x,u),\ldots,f_n(x,u)$. Obviously the curve δ lies on Σ and therefore our proof will be complete if we prove that γ and δ are the same curve. Now, considering that $u = \phi(x)$ is a solution of (1), we have along δ

$$\frac{d\psi}{ds} = \frac{d\phi(\hat{\xi}(s))}{ds}$$

$$= \frac{\partial\phi}{\partial x_1} \cdot \frac{\partial\hat{\xi}_1}{ds} + \ldots + \frac{\partial\phi}{\partial x_n} \cdot \frac{\partial\hat{\xi}_n}{\partial s}$$

$$= f_1(\hat{\xi}(s),\psi(s))\ \frac{\partial\phi}{\partial x_1} + \ldots + f_n(\hat{\xi}(s),\psi(s))\ \frac{\partial\phi}{\partial x_n}$$

$$= f(\hat{\xi}(s),\psi(s)).$$

Therefore, both the curve δ and γ satisfy the same characteristic system and both pass through the same point (x^o, u_o). Because of the uniqueness of the solutions of the Cauchy problem for the characteristic system, it follows that δ and γ are the same curve, q.e.d.

2. The Cauchy Problem in Two Dimensions.- We will now

present the standard Cauchy problem, or initial value problem, for the equation (1). Its importance lies in the fact that it is a well posed problem, that is, it will have a unique solution which has physical significance. Moreover, the Cauchy problem we are about to describe and solve is the simplest of all well posed problems for equation (1), and is therefore the one which yields the most information about the physical sense of (1).

For a better understanding of its content, we state first the problem for only two independent variables. So we let the equation be

$$(3,5) \qquad L[u] \equiv f(x,y,u)\frac{\partial u}{\partial x} + g(x,y,u)\frac{\partial u}{\partial y} - h(x,y,u) = 0 ,$$

where

$$(x,y,u) \in D^3 \subset E^3, \; f,g,h \in C^1(D^3)$$

$$(3,6) \qquad |f| + |g| > 0 , \; (x,y,u) \in D^3 .$$

The corresponding characteristic system is

$$(3,7) \qquad \frac{dx}{f(x,y,u)} = \frac{dy}{g(x,y,u)} = \frac{du}{h(x,y,u)} = ds , \; (x,y,u) \in D^3 .$$

Theorem 3.2. (Cauchy Problem).- _Suppose we are given a curve_ Λ,

$$\Lambda: \begin{cases} x = \alpha(\lambda), \; y = \beta(\lambda), \; u = \kappa(\lambda) \\ \lambda \in I = (\lambda_1, \lambda_2) , \\ \alpha,\beta,\kappa \in C^1(I), \; |\alpha'(\lambda)| + |\beta'(\lambda)| > 0, \; \lambda \in I ; \\ \Lambda \subset D^3 , \end{cases}$$

and such that the closure $\overline{\Lambda}^*$ _of its projection_ Λ^* _on the plane_ $\{0; x,y\}$ _is simple. Moreover, assume that_ Λ^* _satisfies the transversality condition_

$$(3,8) \qquad \begin{vmatrix} \alpha'(\lambda) & \beta'(\lambda) \\ f(\alpha(\lambda),\beta(\lambda),\kappa(\lambda)) & g(\alpha(\lambda),\beta(\lambda),\kappa(\lambda)) \end{vmatrix} \neq 0, \; \lambda \in I.$$

Then there exists a unique integral surface Σ,

Σ: $u = \phi(x,y)$, $(x,y) \in G$, $\Lambda^* \subset G$,

which goes through the curve Λ, that is

$\kappa(\lambda) = \phi(\alpha(\lambda), \beta(\lambda))$, $\lambda \in I$.

Moreover, $\phi(x,y)$ depends continuously on the initial data $\kappa(\lambda)$.

We shall make the statement more precise, and prove:

I) uniqueness, that is, there exists at most one solution Σ through Λ,

II) existence, that is, there exists at least one solution Σ through Λ, and

III) the continuous dependence of φ on the initial data.

Proof of I). We shall first explain what is meant by the uniqueness of Σ. Suppose there exists a second integral surface $u = \hat{\phi}(x,y)$, $(x,y) \in \hat{G}$, $\Lambda^* \subset \hat{G}$, which goes through Λ. That Σ is unique means that for all $(x,y) \in G \cap \hat{G}$, $\phi(x,y) = \hat{\phi}(x,y)$. By Theorem 3.1., if a surface Σ exists it must be given by the integral curves of the characteristic system(7):

$$
\begin{aligned}
x &= \xi(s;\, s_o,\, \alpha(\lambda),\beta(\lambda),\kappa(\lambda)), \quad \xi(s_o) = \alpha(\lambda) \\
y &= \eta(s;\, s_o,\, \alpha(\lambda),\beta(\lambda),\kappa(\lambda)), \quad \eta(s_o) = \beta(\lambda) \qquad (3,9) \\
u &= \zeta(s;\, s_o,\, \alpha(\lambda),\beta(\lambda),\kappa(\lambda)), \quad \zeta(s_o) = \kappa(\lambda),
\end{aligned}
$$

where $\lambda \in I$, and $|s-s_o| < \delta(\lambda)$. The existence of $\delta(\lambda) > 0$ for all pertinent values of λ is insured by the existence theorem of characteristic curves. On the other hand, it is known that the integral curves (9) are unique.

This proves the uniqueness of the solution Σ in the sense given above.

II). In order to prove the existence of G and ϕ, we begin by showing we can solve the first 2 equations of (9) for s and λ as functions of (x,y) for the values of s and λ corresponding to Λ^*, that is, when $s = s_o$ and $\lambda \in I$. It then follows from the continuity argument that there is, in the plane {0; x,y}, a neighborhood of Λ^*, which we take as G, in which ζ will be defined as a continuously differentiable function of (x,y). And, we can set

$$\phi(x,y) = \zeta(s(x,y) \; ; \; s_o, \; \alpha(\lambda(x,y)), \; \beta(\lambda(x,y)), \; \kappa(\lambda(x,y))),$$

$$(3,10) \qquad (x,y) \in G.$$

But now we can solve the first two equations of (9) for (s,λ) as functions of (x,y) for the points $s = s_o$, $\lambda \in I$ if and only if

$$(3,11) \qquad \frac{\partial(\xi,\eta)}{\partial(s,\lambda)} \neq 0 \; , \quad s = s_o \; , \quad \lambda \in I \; .$$

On one hand, for a fixed λ, one has from (7)

$$\frac{\partial\xi}{\partial s} = \frac{dx}{ds} = f(x,y,u)$$

$$\frac{\partial\eta}{\partial s} = \frac{dy}{ds} = g(x,y,u) \; ,$$

and on the other hand, from (9)

$$\xi(s_o) = \alpha(\lambda), \; \eta(s_o) = \beta(\lambda), \; \zeta(s_o) = \kappa(\lambda), \; \lambda \in I.$$

Hence the condition (11) on the Jacobian reduces to the transversality condition (8) that we have assumed. Therefore we have proven the existence of a surface $u = \phi(x,y)$ for $(x,y) \in G$. We call this surface Σ and show that it actually is a solution of equation (5). To this end, we recall that according to a regularity theorem for solutions of a differential system such as (7), the functions ξ, η, ζ are of class C^1 with respect to their arguments s, α, β, κ, and since the functions α, β, κ are of class C^1 with respect to the independent variable λ, we have that ξ, η and ζ are of class C^1 with respect to s and λ. Therefore, there is a tangent plane at every point in which the coordinate curves s and λ are transverse. But, condition (11) expresses precisely that at all points of the domain G, this transversality condition is fulfilled. Hence we have shown that there is a continuous tangent plane to the surface $u = \phi(x,y), (x,y) \in G$.

To prove that $u = \phi(x,y)$ satisfies (5), we could proceed as indicated in the geometric interpretation of ∮ 1., article 2., or considering that for all $(x,y) \in G$, the characteristic curve through $(x,y,\phi(x,y))$ lies on the surface Σ and that therefore along this curve one has

$$p\,dx + q\,dy = du .$$

Hence

$$\frac{dx}{f} = \frac{dy}{g} = \frac{du}{h} = \frac{p\,dx + q\,dy - u}{fp + gq - h} = \frac{0}{L[\,u\,]} ,$$

whence $L[\,u\,] = 0$, q.e.d.

III). To express precisely and simply what it means to say that $\phi(x,y)$ depends continuously on the initial data $\kappa(\lambda)$, in a non-linear problem such as ours, we must first make several definitions and introduce some new conditions.

Let $\hat{\kappa}(\lambda)$, $\lambda \in I$ and $\hat{\kappa} \in C^1(I)$, fulfill the transversality condition (8) and let $\hat{\Lambda} \subset D^3$ be the curve defined by α, β, $\hat{\kappa}$. Moreover, suppose there is an M such that

$$|\kappa(\lambda)| \leqq M\ ,\ |\hat{\kappa}(\lambda)| \leqq M,\ \lambda \in I.$$

Let $u = \hat{\phi}(x,y)$, $(x,y) \in \hat{G}$, be the integral surface of (5) through $\hat{\Lambda}$.

The first two equations of (9) define a mapping H of the pairs (s,λ) onto G. We refer the pairs (s,λ) to a Cartesian coordinate system $\{0;\ s,\lambda\ \}$ and set

$$H^{-1}(G) = J \subset \{0';\ s\ \lambda\}.$$

Then H is a homeomorphism of J onto G. Let Q be any compact set in J such that $\hat{H}(Q) \subset \hat{G}$ where $\hat{H}$ is the same homeomorphism as H, but with $\hat{\kappa}(\lambda)$ instead $\kappa(\lambda)$. Finally we define $\hat{\phi}(x,y)$, $\phi(x,y)$ as functions of (s,λ) by

$$(3,12)\begin{cases} \phi(x,y) = \zeta(s;s_o,\alpha(\lambda),\beta(\lambda),\kappa(\lambda)) = \psi(s,\lambda),(s,\lambda) \in Q, \\ \hat{\phi}(x,y) = \zeta(s;s_o,\alpha(\lambda),\beta(\lambda),\hat{\kappa}(\lambda)) = \hat{\psi}(s,\lambda),(s,\lambda) \in Q. \end{cases}$$

Then to say $\phi(x,y)$ depends continuously on the initial data $\kappa(\lambda)$ means the following:

There is a constant K depending on M and Q, but not on $\hat{\kappa}$, such that

$$(3,13)\qquad |\hat{\psi}(s,\lambda) - \psi(s,\lambda)| \leqq K.|\hat{\kappa}(\lambda) - \kappa(\lambda)|\ ,\ (s,\lambda) \in Q.$$

Actually, according to (12) one has

$$\hat{\psi}(s,\lambda) - \psi(s,\lambda) = \zeta(s;s_o,\alpha(\lambda),\beta(\lambda),\hat{\kappa}(\lambda)) - \zeta(s;\ s_o,\alpha(\lambda),\beta(\lambda),\kappa(\lambda)).$$

Now for fixed λ and variable s, the two terms of the second member represent the two third components of the two vector solutions of system (7). These two solutions pass respectively thru the points

$$\begin{cases} x = \alpha(\lambda),\ y = \beta(\lambda),\ u = \hat{\kappa}(\lambda) \\ x = \alpha(\lambda),\ y = \beta(\lambda),\ u = \kappa(\lambda), \end{cases}$$

in such a way that the distance between the two initial points is

$$| \hat{\kappa}(\lambda) - \kappa(\lambda) |.$$

Now, by a well known estimate of the distance between two solutions for the same s value when one knows the distance between initial values, we have

$$| \zeta(s ; s_o, \alpha, \beta, \hat{\kappa}) - \zeta(s ; s_o, \alpha, \beta, \kappa) | \leq$$

$$\leq | \hat{\kappa}(\lambda) - \kappa(\lambda) | \cdot \exp (A | s-s_o |), \tag{3,14}$$

where A is a Lipschitz constant of the vector (f,g,h) with respect to the variables x,y,u when (x,y,u) varies along the compact set determined by

$$(x,y) \in H(Q), \ | u | \leq M.$$

It is clear that we can estimate $|s - s_o|$ for $(s,\lambda) \in Q$ and the constant A by means of sums of absolute values of the continuous first derivatives of f,g,h with respect to x,y,u. Substituting these majorant values in place of A and $|s - s_o|$ in (14), we arrive at the estimate of (13), q.e.d.

This completes the proof of the theorem.

3. <u>The Cauchy Problem in n-Dimensions.</u>- We will give only the statement since the proof is similar to the 2-dimensional case. The following statement refers to equation (1).

<u>Theorem 3.3</u>. <u>Suppose we are given the</u> (n-1)-<u>dimensional manifold</u> Λ,

$$\Lambda: \begin{cases} x = \alpha(\lambda), \ u = \kappa(\lambda), \ \alpha = (\alpha_1, \ \ldots \ , \alpha_n) \\ \lambda = (\lambda_1, \ \ldots \ , \lambda_{n-1}) \in J, \text{ domain of } E^{n-1} \\ \alpha \ , \kappa \in C^1(J), \\ \Lambda \subset D^{n+1}, \end{cases}$$

such that the closure $\overline{\Lambda}^*$ *of its projection* Λ^* *in the hyperplane* $\{0\ ;\ x\}$ *is simple.*

Moreover, suppose Λ^* *satisfies the transversality condition*

$$(3,15)\qquad \begin{vmatrix} \dfrac{\partial\alpha_1}{\partial\lambda_1} & \cdots\cdots & \dfrac{\partial\alpha_n}{\partial\lambda_1} \\ \vdots & & \\ \dfrac{\partial\alpha_1}{\partial\lambda_{n-1}} & \cdots\cdots & \dfrac{\partial\alpha_n}{\partial\lambda_{n-1}} \\ f_1(\alpha(\lambda),\ \kappa(\lambda)) & \cdots & f_n(\alpha(\lambda),\ \kappa(\lambda)) \end{vmatrix} \neq 0\ ,\ \lambda \in J.$$

Then there exists a unique integral hypersurface Σ *defined by* $u = \phi(x),\ x \in G,$ *which passes through* Λ, *that is*

$$\kappa(\lambda) = \phi(\alpha(\lambda)),\ \lambda \in J\ .$$

Moreover, the hypersurface Σ *depends continuously on the only initial data* $\kappa(\lambda)$.

This hypersurface is parametrically given by the solutions of the system (1,5b),

$$(3,16)\qquad \Sigma: \begin{cases} x = \xi(s;s_o,\alpha(\lambda),\kappa(\lambda)) \equiv \xi(s), & \xi(s_o) = \alpha(\lambda) \\ u = \zeta(s;s_o,\alpha(\lambda),\kappa(\lambda)) \equiv \zeta(s), & \zeta(s_o) = \kappa(\lambda) \end{cases}$$

$$\lambda \in J\ ,\ |\ s - s_o| < \delta(\lambda) > 0.$$

The transversality condition (15) says

$$\frac{\partial(\xi)}{\partial(\lambda, s)} \neq 0,\ s = s_o\ ,\ \lambda \in J\ ,$$

which is to say one can solve the first equation of (16) for s and λ, and substituting $s(x)$ and $\lambda(x)$ into the last equation of (16) obtain the integral hypersurface $u = \phi(x)$.

4. *Examples*.- For a better comprehension of the

preceeding sections 2. and 3., we recommend the study of examples, and, in particular, the following two which are interesting in themselves.

a) Study the complete linear partial differential equation of first order with 2 independent variables,

$$f(x,y)\frac{\partial u}{\partial x} + g(x,y)\frac{\partial u}{\partial y} - h(x,y)\cdot u = k(x,y),$$

$$f,g,h,k \in C^1(D^2) , D^2 \subset \{0 ; x,y\} .$$

$$|f| + |g| > 0 , (x,y) \in D^2.$$

Prove there exists a first integral of the corresponding characteristic system which depends only on x,y. Call it w(x,y) and carry out the explicit calculations of the general integral only by two quadratures. Given the general integral, find the equation.

Consider the geometric interpretation of the family of curves w(x,y) = constant which are base characteristic curves in the plane {0 ; x,y}. Compare characteristic curves through points of the same straight line parallel to the u-axis. What happens tó these curves when h = 0?

Given w(x,y), indicate the solution of the Cauchy problem. Refer to exercise 1. of this section.

b) Study the equation

$$u \cdot \frac{\partial u}{\partial x} - \frac{\partial u}{\partial y} = 0 .$$

Find explicitly the system of characteristic curves and construct integral surfaces.

Show that there is no curve Λ^* which satisfies the transversality condition at none of its points, and is not the projection of a characteristic curve. The construction of integral surfaces through a curve Λ, whose projection is a base characteristic curve, is not excluded even if Λ is not a characteristic curve.

Exercises

1. Prove that for the semilinear equation

$$f(x,y) \cdot \frac{\partial u}{\partial x} + g(x,y) \cdot \frac{\partial u}{\partial y} = h(x,y,u) ,$$

there is always a first integral $w_1(x,y)$ depending only on x and y of its characteristic system. Moreover, the set of all base characteristic curves depends on only one parameter.

2. Consider the following linear equation:

$$f(x,y) \cdot \frac{\partial u}{\partial x} + g(x,y) \frac{\partial u}{\partial y} + k(x,y) \cdot u = m(x,y) .$$

Prove the following

a) If one knows two integral surfaces, then one can find a 1-parameter system of solutions; if one knows three linearly independent solutions, then one can obtain a two parameter system of solutions.

b) If one knows a first integral $w(x,y)$ of the characteristic system, then one can obtain the general integral of the given equation through two quadratures. The general integral is of the form

$$u = \sigma(x,y) + \tau(x,y) \cdot f(w_1(x,y)),$$

where τ and σ are determined and f is arbitrary. [Note the analogy between this general solution and the general solution of an ordinary linear differential equation of first order.]

c) If $m(x) \equiv 0$ and $v(x,y) = 0$, $u = z(x,y)$ are the equations of a characteristic curve, then any other characteristic curve whose projection is $v(x,y) = 0$ has as equations

$$v(x,y) = 0, \quad u = c \cdot z(x,y) \text{ , c a constant.}$$

d) If $k(x,y) \equiv 0$, and $v(x,y) = 0$, $u = z(x,y)$ are the equations of a characteristic curve, then any other characteristic curve whose projection is $v(x,y) = 0$ has as equations

$$v(x,y) = 0 \ , \ u = z(x,y) + c \ , \ c \text{ constant.}$$

e) If $m(x) \equiv 0 \equiv k(x)$, then the characteristic curves have equations of the form

$$v(x,y) = 0, \ u = c \ , \ c \text{ arbitrary constant.}$$

Moreover, in this case, study the relation between the general integral $u = f(w(x,y))$ and the integrating factor of the characteristic equation

$$f \, dy - g \, dx = 0 \ .$$

3. Let $w_1(x,y,u)$, $w_2(x,y,u)$ be two functionally independent first integrals of the characteristic system of the equation

$$f(x,y,u) \frac{\partial u}{\partial x} + g(x,y,u) \frac{\partial u}{\partial y} = h(x,y,u) \ ;$$

and let

$$x = \alpha(\lambda) \ , \ y = \beta(\lambda) \ , \ u = \gamma(\lambda)$$

be the equations of an initial transverse curve.

Show that if $\sigma(c_1,c_2) = 0$ is the result of eliminating x,y,u,λ between the equations of the initial curve and the two relations

$$w_1(x,y,u) = c_1 \ , \ w_2(x,y,u) = c_2 \ ,$$

then

$$\sigma(w_1(x,y,u) \ , \ w_2(x,y,u)) = 0$$

is the equation of the integral surface through the given initial curve. Note that this method may be simpler than that in the text.

4. Given the equation

$$(y-u)\frac{\partial u}{\partial x} + (x-y)\frac{\partial u}{\partial y} = u - x ,$$

find the integral surface which passes through the parabola

$$y - 1 = 0 , \quad u - x^2 = 0 .$$

Observe what happens at the point of the initial curve where $x = -1/2$.

5. Given the equation

$$\begin{vmatrix} \frac{\partial u}{\partial x} & \frac{\partial u}{\partial y} & -1 \\ yu & xu & xy \\ x & y & u \end{vmatrix} = 0 ,$$

find the integral surface through the curve

$$x = t,\ y = t,\ u = \frac{1}{t^2} .$$

6. Given the equation

$$(x+y)\frac{\partial u}{\partial x} + (u-x)\frac{\partial u}{\partial y} = y + u,$$

find the general integral and the integral surface through the x-axis.

7. We can give a more general definition of an integral surface as follows. One says that the surface

$$x = \xi(s,\lambda) , \quad y = \eta(s,\lambda) , \quad u = \zeta(s,\lambda) , \quad (s,\lambda) \in H \subset E^2$$

is a solution (in parametric form) of the equation

$$f(x,y,u) \cdot u_x + g(x,y,u) \cdot u_y = h(x,y,u)$$

if

1) $\xi, \eta, \zeta \in C^1(H)$ and,

2) $f(\xi(s,\lambda), \eta(s,\lambda), \zeta(s,\lambda)) \frac{\partial(\eta,\zeta)}{\partial(s,\lambda)} + g \frac{\partial(\zeta,\xi)}{\partial(s,\lambda)} + h \frac{\partial(\xi,\eta)}{\partial(s,\lambda)} = 0,$

for $(s,\lambda) \in H$.

Suppose we are also given an intial curve

$x = \alpha(\lambda),\ y = \beta(\lambda),\ u = \kappa(\lambda)$

satisfying

$|\alpha'| + |\beta'| + |\kappa'| > 0$,

which is somehow a weaker condition than the corresponding one on the case of a solution in explicit form.

Prove that there exists a unique integral surface in parametric form through the given curve. In place of conditions (6) and (8), it is enough to require

$|f| + |g| + |h| > 0$,

and that the matrix

$$\begin{pmatrix} \alpha'(\lambda) & \beta'(\lambda) & \kappa'(\lambda) \\ f(\alpha(\lambda), \beta(\lambda), \kappa(\lambda)) & g & h \end{pmatrix}$$

has rank two.

Note: The generalization consists in considering both simultaneously and **separately** the two equations

$f \cdot y_x - g + h\, y_u = 0$, $-f + g\, x_y + h\, x_u = 0$

and the given equation.

8. Suppose we are given an initial curve as in the preceeding exercise, and assume that the mentioned matrix has rank one.

Prove that the curve is a characteristic curve. (See Theorem 8.3).

9. Given the equation

$$(xy - u) \frac{\partial u}{\partial x} + (y^2 - 1) \frac{\partial u}{\partial y} = uy - x ,$$

find the integral surface through the curve

$$y = 0 , x^2 - u^2 = 1 .$$

Also, find the integral surface through the curve

$$x^2 + y^2 = 1 , u = 0 .$$

[The second one does not exist in explicit form, but does exist in parametric form and coincides with the first integral surface.]

10. Suppose we are given the equation

$$u u_x - u_y = 0 .$$

(especially interesting when one looks for non-regular solutions, cf Courant-Hilbert, App. 2 to Ch. II).

a) Find the general integral of the characteristic curves.

$$[u = u_o , y = y_o - \frac{(x-x_o)}{u_o} .]$$

b) Find the integral surface through the (continuously differentiable) curve

$$y = 0 , u = \kappa(x) ;$$

and verify that this surface always exists (in explicit form).

$[u = \kappa(x + yu)]$.

c) Prove that if one has

$$\kappa(x_o) = 0 , |\kappa'(x)| \geqq M \text{ for } |x - x_o| \leqq \frac{1}{M} ,$$

then the integral surface fails to exist at a point (x_o, y_o) such that $|y_o| \leqq 1/M$.

d) Show that if

$$\kappa(x_o) = 0 \ , \ \kappa'(x_o) = \infty \ ,$$

then there is no integral surface in a neighborhood of $x = x_o$, $y = 0$.

c) Prove that there exists no integral surface through the continuously differentiable curve

$$x = \alpha(\lambda), \ y = \beta(\lambda), \ u = \kappa(\lambda)$$

in a neighborhood of a point λ_o where the transversality condition is not satisfied, that is, when

$$\alpha'(\lambda_o) + \beta'(\lambda_o) \cdot \kappa(\lambda_o) = 0 \ ,$$

except in the case where one also has $\kappa'(\lambda_o) = 0$, which is to say the given curve is tangent at λ_o to the characteristic curve through $(\alpha(\lambda_o), \beta(\lambda_o), \kappa(\lambda_o))$.

Chapter II

GENERAL EQUATION WITH TWO INDEPENDENT VARIABLES

§ 4. Monge Curves

1. The Standard Conditions.- We are now going to study the equation in general form, that is, not quasi-linear, but, for the present, with only two independent variables.

Suppose we are given the equation

(4,1a) $$F(x,y,u,p,q) = 0 \ , \ p = \frac{\partial u}{\partial x} \ , \ q = \frac{\partial u}{\partial y}$$

(4,1b) $$F \in C^2(D^5) \ , \ |F_p| + |F_q| \neq 0 \ ,$$

$$(x,y,u,p,q) \in D^5 \subset E^5 \ , \ (x,y,u) \in D^3 \subset E^3 .$$

The domain D^3 is of course the projection of D^5 on the space of its first three coordinates.

We say that $u = \phi(x,y)$ is a solution or surface integral of (1) if

1) there exists a domain $G \subset \{0 \ ; \ x,y\}$ where

(4,2a) $$\phi \in C^2(G),$$

2) and one has

(4,2b) $$F(x,y,\phi(x,y), \ \phi_x(x,y), \ \phi_y(x,y)) = 0 \ , \ (x,y) \in G.$$

It is understood that condition (2) implies that for each $(x,y) \in G$,

$$(x,y,\phi(x,y), \ \phi_x(x,y), \ \phi_y(x,y)) \in D^5 .$$

It is assumed that both F and ϕ are real. That is, we assume that $F(x,y,u,p,q) = 0$ defines an open real 4-dimensional manifold in the Euclidean space $E^5 = \{0; x,y,u,p,q\}$.

The condition $\phi \in C^2(G)$ is not natural, since the nature of the problem seems only to require of the integral surface that it has a continuous tangent plane at all of its points, as was assumed in the case of a quasi-linear equation. But, the proofs we will give require the existence of dp and dq and hence we must assume that p and q are differentiable. It may be interesting to point out that some of the most important results we shall obtain (unique solution of the Cauchy problem and generation of surface integrals by characteristic strips) are also valid assuming only $\phi \in C^1$.

We should also remark that our results will be of a local nature.

2. Definitions and Geometric Interpretation.- We define a surface differential element of first order, or simply, element, to be a vector or ordered set of five numbers $(x_o, y_o, u_o, p_o, q_o) = E^o$, where $E^o \in D^5$.

The set of elements of D^5 depends on 5 parameters and the primary meaning of (1) is that the differential equations determine a subset of real elements depending on 4 arbitrary, real parameters, which we will call integral elements and are those which satisfy equation (1). The geometric interpretation of an element $E^o = (x_o, y_o, u_o, p_o, q_o)$ as a plane characterized by its supporting point $M_o = (x_o, y_o, u_o) \in D^3$ and its normal $(p_o, q_o, -1)$ is obvious. Consider the set of elements whose supporting point is the fixed point M_o. Then the equation

(4,3) $$F(M_o, p, q) = 0 ,$$

determines a set of planes through M_o whose normals are given by $(p,q,-1)$ where p,q satisfy (3). These planes envelop a cone which corresponds to $M_o \in D^3$ through (3).

These cones are called Monge cones and may be considered as differential elements of conical points defined for each point M_o by the corresponding equation (3).

In the case of a quasi-linear equation, the cone (3) is the set of planes through the vector which we associated to each point. Now we can picture D^3 as a field of Monge cones, one at each point. It is clear that for a surface of class two to be an integral surface, it is necessary and sufficient that the tangent plane at each point also be tangent to the Monge cone of that point along a generator.

We define a Monge curve (also called focal curve) to be one with a continuous tangent such that the tangent at each point is a generator of the Monge cone (of that point). Since the Monge cone at any point consists of infinitely many generators depending on one arbitrary parameter, it is obvious that there are equally many Monge curves at any point of D^3. These curves may, so to say, branch off at each point, provided that the direction of the tangent (at each point) coincides with that of a generator of the Monge cone at the point.

We call a surface strip, or simply, strip, the set consisting of a curve, called the supporting curve

$$x = \alpha(\lambda),\ y = \beta(\lambda),\ u = \kappa(\lambda),\ \lambda \in I = (\lambda_o,\lambda_1)$$

$$\alpha,\beta,\kappa \in C^1(I),$$

and a normal $(p,q,-1)$ to the tangent of the supporting curve

$$p = \delta(\lambda)\ ,\ q = \eta(\lambda),\ \lambda \in I,\ \delta,\ \eta \in C^1(I)$$

where

$$(4,4)\quad \delta(\lambda)\cdot\alpha'(\lambda)+\eta(\lambda)\cdot\beta'(\lambda)-\kappa'(\lambda)=0\ ,\ \lambda\in I.$$

The last condition is called the strip condition because it is a necessary and sufficient condition for the 1-parameter system of elements $(\alpha(\lambda), \beta(\lambda), \kappa(\lambda), \delta(\lambda), \eta(\lambda)) \equiv E(\lambda)$ to constitute a strip. This condition expresses that in fact the normal whose direction parameters are $(p = \delta(\lambda),$ $q = \eta(\lambda), -1)$ is orthogonal to the tangent of the supporting curve, whose direction parameters are

$$dx = \alpha'(\lambda)d\lambda\ ,\ dy = \beta'(\lambda)d\lambda\ ,\ du = \kappa'(\lambda)d\lambda\ .$$

We say that a strip is an integral strip if all of its elements are integral elements. A strip is called a Monge strip if it is integral and its supporting curve is a Monge curve. It is obvious that a Monge curve can be completed to a Monge strip.

In fact, if μ is a Monge curve and $M_o \equiv (x_o, y_o, u_o) \in \mu$, we have only to take for p_o and q_o those values which correspond to a tangent plane to the Monge cone at M_o along the generator which is the straight line tangent to μ at M_o. Moreover, it follows that given the focal curve μ and an integral element $E^o = (M_o, p_o, q_o)$, the focal strip supported by μ is unique.

3. Differential Systems of Monge Curves.- Let $M_o =$ $(x_o, y_o, u_o) \in D^3$ and let μ be a Monge curve through M_o,

$$(4,5)\quad \begin{aligned} &\mu : x = \xi(s),\ y = \eta(s),\ u = \zeta(s) \\ &\quad s \in I_s = \{s \mid -\delta_1 < s - s_o < \delta_2\} \\ &\quad \xi(s_o) = x_o\ ,\ \eta(s_o) = y_o\ ,\ \zeta(s_o) = u_o\ , \\ &\qquad \xi,\ \eta,\ \zeta \in C^1(I_s). \end{aligned}$$

The Monge cone K_o at M_o is the cone enveloped by the planes

(4,6a) $$(X - x_o)p + (Y - y_o)q - (U - u_o) = 0 ,$$

where X, Y, U are the present coordinates of the plane and p,q are related by the equation

(4,6b) $$F(x_o,y_o,u_o,p,q) = 0 .$$

Now, because of (1), either F_p or F_q must be different from zero. Suppose $F_p \neq 0$. Then fixing an integral element (M_o,p_o,q_o) and letting q vary, (6) defines p as a function of q. The generators of the cone K_o are given as intersections of the plane (6a) and the plane

$$(X - x_o) \frac{dp}{dq} + (Y - y_o) = 0 ,$$

obtained by taking derivatives of (6a), where

$$F_p(x_o,y_o,u_o,p,q) \frac{dp}{dq} + F_p = 0 .$$

Hence the equations of the generators of the cone K_o are, together with (6b),

(4,7) $$\frac{X - x_o}{F_p(x_o,y_o,u_o,p,q)} = \frac{Y - y_o}{F_q} = \frac{U - u_o}{p\,F_p + q\,F_q} ,$$

Therefore, for each value of q there corresponds a generator of the Monge cone K_o.

From (7), one sees that the direction parameters of the generators are $(F_p,F_q,pF_p + qF_q)$. On the other hand, the direction parameters of the tangent to the Monge curve μ are dx, dy, du. Since both must coincide, for the points (x,y,u) of a Monge curve, we have that

(4,8a) $$F(x,y,u,p,q) = 0$$

$$(4,8b)\qquad \frac{dx}{F_p(x,y,u,p,q)} = \frac{dy}{F_q} = \frac{du}{pF_p + qF_q} = ds ,$$

are the differential equations of the Monge curves.

In the equations (8), the 5 variables (x,y,u,p,q) are related in such a way that instead of considering only the curve μ described by the point (x,y,u), it will be more convenient, due to symmetry and for the sake of simplicity, to consider (x,y,u,p,q) as a 1-parameter system of elements depending on the auxilliary variable s. Therefore, we complete the equation (5) of μ with

$$(4,9)\qquad \begin{aligned} &p = \pi(s),\ q = \rho(s),\ s \in I_s\ ;\ \pi,\ \rho \in C^1(I_s) ,\\ &\pi(s_o) = p_o,\ \rho(s_o) = q_o,\ F(x_o,y_o,u_o,p_o,q_o) = 0 . \end{aligned}$$

In this way, as s varies, (5) and (9) define a 1-parameter system of elements, and the curve μ will be a Monge curve if and only if equations (8), which now take the form

$$(4,10a)\qquad F(\xi(s),\ \eta(s),\ \zeta(s),\ \pi(s),\ \rho(s)) = 0 ,\ s \in I_s ,$$

$$(4,10b)\qquad \frac{dx}{F_p(x,y,u,\pi(s),\rho(s))} = \frac{dy}{F_q} = \frac{du}{\pi(s).F_p + \rho(s).F_q} = ds,$$

are satisfied.

It is important to note that the 1-parameter system of elements defined by (5) and (9) satisfying (10b) form a strip. Actually, from (10b), it follows that

$$\pi(s) \cdot \xi'(s) - \rho(s) \cdot \eta'(s) - \zeta'(s) = 0 ,$$

which is the stip condition. Therefore, if $\mu$is a Monge curve, since by virtue of (10a) all the elements are integral, it follows that (5) and (9) define a Monge strip.

We summarize these results that we obtained in the following assertion.

Assertion 4.1 Let $E^o = (x_o, y_o, u_o, p_o, q_o) \in D^5$ be an integral element, $F_p(E^o) \neq 0$, and $D_o^5 \subset D^5$ a neighborhood of E^o.

1) In order that the 1-parameter system of integral elements $E(s)$,

$$(4,11) \quad \begin{aligned} &E(s) : x = \xi(s),\ y = \eta(s),\ u = \zeta(s),\ p = \pi(s),\ q = \rho(s); \\ &s \in I_s = \{s| - \delta_1 < s - s_o < \delta_2\},\ E(s_o) = E^o ; \\ &\xi,\ \eta,\ \zeta, \pi, \rho \in C^2(I_s), \end{aligned}$$

be a Monge strip it is necessary and sufficient that it satisfy $F = 0$ and the differential system of Monge curves, that is, we must have

$$(4,12) \begin{cases} F(x,y,u,p,q) = 0 \\ \dfrac{dx}{F_p(x,y,u,\pi(s),\rho(s))} = \dfrac{dy}{F_q} = \dfrac{du}{\pi(s)\cdot F_p + \rho(s)\cdot F_q} = ds, \\ s \in I_s . \end{cases}$$

2. Let $\rho(s)$ be an arbitrary function as in 1). And let $p = \pi(s;\ x,y,u)$ be the function implicitly defined by $F(x,y,u,p,\rho(s)) = 0$. Then the unique solution through $(s_o;\ x_o, y_o, u_o)$

$$\begin{cases} x = \xi(s\ ;\ s_o,\ E^o) \equiv \xi(s),\ \xi(s_o) = x_o \\ y = \eta(s\ ;\ s_o,\ E^o) \equiv \eta(s),\ \eta(s_o) = y_o \\ u = \zeta(s\ ;\ s_o,\ E^o) \equiv \zeta(s),\ \zeta(s_o) = u_o\ , \end{cases}$$

of the differential system

$$\frac{dx}{F_p(x,y,u,\pi(s;\ x,y,u),\rho(s))} = \frac{dy}{F_q} = \frac{du}{\pi(s;x,y,u)\cdot F_p+\rho(s)\cdot F_q} = ds,$$

is a Monge curve which is completed to a Monge strip by the addition of the functions

$$p = \pi(s;x,y,u), = q = \rho(s).$$

Therefore, given E^o and $\rho(s)$, the Monge strip is unique.

The second part is an immediate consequence of the first, and manifestly shows the amount of indetermination of system (12).

We conclude the study of Monge strips proving the following assertion which shows that if there are integral surfaces of (1), they may be considered as generated by juxtaposition of Monge strips.

Assertion 4.2. Let $u = \phi(x,y), (x,y) \in G$, be an integral surface of $F = 0$. Let $(x_o,y_o) \in G$, $\phi(x_o,y_o) = u_o$. Then the 1-parameter system of elements

$$(4,13) \quad \begin{cases} x = \xi(s\ ;\ s_o,\ x_o,\ y_o,\ u_o) \\ y = \eta(s\ ;\ s_o,\ x_o,\ y_o,\ u_o) \\ u = \zeta(s\ ;\ s_o,\ x_o,\ y_o,\ u_o) \\ p = \phi_x(x,y),\ q = \phi_y(x,y),\ s \in I_s\ , \end{cases}$$

where ξ, η, ζ are the unique solution of the system

$$(4,14) \quad \frac{dx}{F_p(x,y,u,\phi_x(x,y),\phi_y(x,y))} = \frac{dy}{F_q} = \frac{du}{\phi_x\cdot F_p+\phi_y\cdot F_q} = ds\ ,$$

is a Monge strip lying in the integral surface.

As an immediate consequence of this assertion, it follows that if we construct a curve Λ on the integral surface which is transverse to the Monge strips that pass through the elements of Λ, then these strips juxtaposed generate the integral surface and they may be defined as originating at the points (x_o, y_o, u_o) of the initial transverse curve Λ.

Proof. Applying assertion 4.1, one sees that (13) defines a Monge strip. To prove that this strip lies in the given surface, it is enough to prove that the curve defined by the first three equations of (13) lies in the given surface, since the normal to the strip is the same as the normal to the surface. To prove that this curve lies in the surface, it is enough to consider that from equations (14) it follows that $w(x,y,u) = u-\phi(x,y)$ is a first integral of the same system (14).

Observe that in general, a sufficiently small part of any surface may be considered as generated by a 1-parameter system of Monge curves. However, the Monge strips supported by these curves will not constitute a surface unless the surface is an integral surface. That is to say, a 1-parameter system of Monge strips does not necessarily constitute a system of juxtaposed strips. On the contrary, in general they will not fit together and, in fact, form only a venetian blind type object.

5. The Cauchy Problem.

1. The Standard Problem.- In the preceeding section we have studied the properties of Monge strips and have seen that if integral surfaces exist, they must be generated by Monge strips, each fitting together with its neighbor to form a surface and not a venetian blind type object. We should not forget that what we are really interested in, is the existence of integral surfaces. In this section we propose and solve a problem for equation (4,1) which determines an integral surface which is unique and depends continuously on the data. We now propose this problem.

Statement of Cauchy Problem. Let Λ be a supporting curve of an integral strip $E^o(r)$ and suppose the following regularity conditions are satisfied:

$$(5,1)\quad \Lambda: \begin{cases} x = \alpha(r),\ y = \beta(r),\ u = \chi(r),\ r \in I_r = (a,b)\ ; \\ \alpha,\beta,\chi \in C^2(I_r)\ ;\ |a'| + |\beta'| > 0,\ r \in I_r\ ; \end{cases}$$

$$(5,2)\quad E^o(r): \begin{cases} \Lambda,\ p = \sigma(r),\ q = \tau(r),\ r \in I_r\ ; \\ p,\ q \in C^1(I_r)\ ; \end{cases}$$

The functions p and q are not arbitrary and, in fact, must satisfy the conditions

$$F(\alpha(r),\beta(r),\chi(r),\sigma(r),\tau(r)) = 0$$

$$(5,3)\quad \sigma(r)\cdot\alpha'(r) + \tau(r)\cdot\beta'(r) - \chi'(r) = 0,\ r \in I_r\ .$$

The first of these conditions assures that all elements of $E^o(r)$ are integral, and the second, that they constitute a strip.

Moreover, assume that the closure of the projection Λ^* of Λ on the plane $\{0 \,;\, x,y\}$ is simple and satisfies the transversality condition

$$(5,4) \qquad T(r) \equiv \begin{vmatrix} F_p(\alpha(r),\beta(r),\chi(r),\sigma(r),\tau(r)) & F_q \\ \alpha'(r) & \beta'(r) \end{vmatrix} \neq 0, \; r \in I_r.$$

Observe that although the transversality condition is predicated of Λ^*, its definition depends on the strip $E^o(r)$ in all components. If one arbitrarily gives the curve Λ by (1) and an initial integral element $E^o(r)$, then the equations (3) will in general uniquely determine $\sigma(r)$ and $\tau(r)$; saying, in general, we mean precisely if and only if the transversality condition (4) is satisfied, since this is a necessary and sufficient condition that (3) may be solved for σ and τ when one has previously fixed an initial element.

We may now state the Cauchy problem for equation (4,1). It consists in constructing an integral surface $u = \phi(x,y)$, $(x,y) \in G$, that contains the strip $E^o(r)$, that is, one has

$$\Lambda^* \subset G \;;\; \chi(r) = \phi(\alpha(r),\beta(r)), \sigma(r) = \phi_x(\alpha(r),\beta(r)),$$

$$(5,5) \qquad \tau(r) = \phi_y(\alpha(r),\beta(r)) \;,\; r \in I_r \;.$$

We have seen that the integral surface must be generated by Monge strips, each fitting together with its neighbor. In order to characterize this particular class of Monge strips, which we call characteristic strips, we proceed as in the quasi-linear case. We start with an integral surface and look for conditions which determine the class of characteristic strips. Then, by means of characteristic strips and beginning with an initial integral strip given by (2), we shall construct a surface which can be shown to be an integral surface.

2. Characteristic Strips.- Suppose that in the first part of Assertion 4.1., the initial element depends on a parameter $r \in I_r$. Then instead of a strip $E(s)$, we will have a 1-parameter system of strips which we may represent by $E(s,r)$, or better, by $E(s,r) \equiv E(s\ ;\ s_o, E^o(r))$ to show clearly that for $s = s_o$ we have the initial integral strip $E^o(r)$, $r \in I_r$.

Then $E(s,r)$ represents a 2-parameter system of elements, and it is clear that the first three components $x = \xi(s,r)$, $y = \eta(s,r)$ and $u = \zeta(s,r)$, when s and r vary, are the equations of a surface. The necessary and sufficient condition that, with r fixed, $E(s,r)$ be a Monge strip, is that they satisfy (4,12). In the case $E(s,r)$ satisfies (4,12), all of the elements of $E(s,r)$ are integral, and, for fixed r, constitute an integral strip as s varies. Our problem consists in showing that for fixed s, one also has that $E(s,r)$, when r varies, generates a strip, and not merely a 1-parameter system of elements which although integral, do not fit together; the only result we have thus far is that $E(s_o,r)$, when r varies over I_r, does form an integral strip, namely the initial strip $E^o(r)$.

Suppose now that the $E(s,r)$ actually are elements of an integral surface Σ: $u = \phi(x,y), (x,y) \in G$, passing through Λ in such a way that one has

$$(5,6) \qquad p \equiv \sigma(s,r) = \phi_x(x,y), \quad q \equiv \tau(s,r) = \phi_y(x,y).$$

Remark. In (6) we may assume that s and r vary over a domain J of the plane $\{0; s,r\}$ such that

$$(5,7a) \quad \{(s,r)|\ r \in I_r\ ,\ -\delta_1(r) < s - s_o < \delta_2(r)\} = J,$$

where δ_1, δ_2 are positive, and so small that the mapping H,

$$(5,7b) \quad H : (s,r) \in J \rightarrow (x,y) \in G\ , \begin{cases} x = \xi(s,r) \\ y = \eta(s,r) \end{cases} ,$$

is a homeomorphism. Then G is the image of J through H and one has $\Lambda^* \subset G$.

To prove the remark, it is enough to prove that the jacobian $\frac{\partial(\xi,\eta)}{\partial(s,r)} \neq 0$ for $r \in I_r$, $s = s_o$. Now actually, for $s = s_o$, one has $x = \alpha(r)$, $y = \beta(r)$, and therefore

$$\frac{\partial\xi}{\partial r} = \alpha'(r)\ ,\ \frac{\partial\eta}{\partial r} = \beta'(r),\ r \in I_r\ ,\ s = s_o\ .$$

Moreover, for fixed r, the Monge strips E(s,r) satisfy condition (4,14); therefore, for fixed r and varying s, we have

$$\frac{\partial\xi}{\partial s} = F_p(x,y,u,\phi_x,\phi_y),\quad \frac{\partial\eta}{\partial s} = F_q\ ;$$

but for $s = s_o$, one has

$$F_p(x,y,u,\phi_x,\phi_y) = F_p(\alpha(r),\ \beta(r),\ \chi(r),\ \sigma(r),\ \tau(r)),$$

$$F_q = F_q(E^o(r)).$$

Hence, it follows, by virtue of the transversality condition (4), that

$$\left(\frac{\partial(\xi,\eta)}{\partial(x,r)}\right)_{s = s_o} = T(r) \neq 0, \; r \in I_r .$$

Now, if the Jacobian is different from zero for $s = s_o$, $r \in I_r$, by the usual continuity argument, it is also different from zero in a neighborhood of this segment in the plane $\{0 \; ; \; s,r\}$. This neighborhood we take as G, and it is clear that all statements of the remark are verified, q.e.d.

We proceed now to a complete characterization of the Monge strips which lie on an integral surface $u = \phi(x,y)$, $(x,y) \in G$.

Since one has

$$F(x,y,\phi(x,y), \; \phi_x(x,y), \; \phi_y(x,y)) = 0 ,$$

the simplest thing to insist on is that

$$(5,8) \quad \begin{cases} \dfrac{\partial F}{\partial x} = F_x + F_u \; p + F_p \dfrac{\partial p}{\partial x} + F_q \dfrac{\partial q}{\partial x} = 0 \\[2ex] \dfrac{\partial F}{\partial y} = F_y + F_u \; q + F_p \dfrac{\partial p}{\partial y} + F_q \dfrac{\partial q}{\partial y} = 0, \quad (x,y) \in G, \end{cases}$$

be satisfied. Note the notation: $\frac{\partial F}{\partial x} \equiv \left(\frac{\partial F}{\partial x}\right)_y$ means the partial derivative of F with respect to x when the only independent variables are x and y: while F_x means the partial derivative of F with respect to x when the independent variables are x,y,u,p,q. In this same manner, we give the meaning of

$$\left(\frac{\partial p}{\partial s}\right)_r \equiv \frac{\partial p}{\partial s} , \qquad \left(\frac{\partial q}{\partial s}\right)_r \equiv \frac{\partial q}{\partial s} .$$

Considering that $\dfrac{\partial q}{\partial x} = \dfrac{\partial^2 \phi}{\partial y \, \partial x} = \dfrac{\partial p}{\partial y}$, and considering condition (4,14), from the first of (8) it follows, for fixed r and varying only s, that one has

$$(5,9a) \qquad -F_x - p\, F_u = \frac{\partial p}{\partial x} \cdot \frac{\partial x}{\partial s} + \frac{\partial p}{\partial y} \cdot \frac{\partial y}{\partial s} = \frac{\partial p}{\partial s} ,$$

and analagously, from the second part of (8), one has

$$(5,9b) \qquad \frac{\partial q}{\partial s} = - F_y - q\, F_u .$$

Therefore, the elements $E(s\, ;\, s_o,\, E^o(r))$, when they are elements of an integral surface and are considered for fixed r, must satisfy the following ordinary differential system which is obtained by adjoining (4,14) to (9);

$$\frac{dx}{F_p(x,y,u,p,q)} = \frac{dy}{F_q} = \frac{du}{p\, F_p + q\, F_q} = \frac{dp}{-F_x - p\, F_u} =$$

$$(5,10) \qquad = \frac{dq}{-F_y - q\, F_u} = ds .$$

This differential system is called the <u>characteristic system</u> of the equation $F = 0$. It is important because it sufficiently characterizes, by five equations, the five unknown functions of the strips which lie on integral surfaces, and also because it may be written directly, knowing only the equation $F = 0$.

Given the initial element $E^* \equiv (x^*,y^*,u^*,p^*,q^*) \in D^5$, the system which results from deleting the last equation of (10) is an ordinary differential system of fourth order which can always be put in normal form and is regular if one takes x or y as an independent variable. Hence, considering s as an auxiliary parameter, there exists a unique path solution of (10), $E(s\, ;\, s_o,E^*)$ such that for $s = s_o$, one has $E(s_o;\, s_o,E^*) = E^*$.

The system (10) is autonomous, hence allowing us to introduce the auxiliary integration constant s_o in such a way that for any value $s_o \in E$, all the unknown functions are functions of $s - s_o$.

Of course, any path solution of (10), for instance $E(s\ ;\ s_o, E^*)$, is a 1-parameter system of elements constituting a strip. Consequently, instead of path solutions of (10), we may speak of strip solutions. Not considering s_o, they depend on 4 essentially arbitrary integration constants.

Let us verify that $F(x,y,u,p,q)$ is a first integral of the system. From (10), it follows

$$\frac{dx}{F_p} = \frac{dy}{F_q} = \frac{F_p dx + F_q dy + F_u du + F_p dp + F_q dq}{0} =$$

$$= \frac{dF}{0} = ds. \tag{5,11}$$

Therefore, as s varies along a strip solution of (10), F remains constant. It follows that the strip solution $E(s\ ;\ s_o, E^*)$ will be an integral strip of $F = 0$ if and only if the initial element E^* is an integral element of $F = 0$, that is, if and only if $F(E^*) = 0$.

The strip solutions of (10) which are integral strips of $F = 0$ are called <u>characteristic strips</u>, and their supporting curves are called <u>characteristic curves</u>; since the initial element must be integral, it follows that they depend on three arbitrary constants.

3. <u>Uniqueness of the Solution of the Cauchy Problem</u>.- We will now show that if there are integral surfaces of $F = 0$,

they are generated by a 1-parameter system of characteristic strips. This is the content of the following theorem.

Theorem 5.1. Let Σ be an integral surface of $F = 0$ and E^* an element of Σ:

(5,12) $\quad \Sigma: u = \phi(x,y), \ (x,y) \in G$

(5.13) $\quad E^* \equiv (x^*,y^*,\phi(x^*,y^*),\phi_x(x^*,y^*),\phi_y(x^*,y^*))$.

Let $E(s;\ s_o,E^*)$ be a characteristic strip such that $E(s_o;s_o,E^*) = E^*$ and such that $\gamma^* \subset G$, where γ^* is the projection on the plane $\{0;\ x,y\}$ of its supporting curve γ. Then the strip $E(s)$ lies in Σ.

Proof. Set $E(s;\ s_o,E^*) \equiv E(s) \equiv (\xi(s),\eta(s),\zeta(s),\pi(s),\rho(s))$. The strip $E(s)$ is the unique solution of the characteristic system (10) such that $E(s_o) = E^*$. We need only show that $\gamma \subset \Sigma$ where $\gamma = (\xi(s),\eta(s),\zeta(s))$, that is to say

$$\zeta(s) = \phi(\xi(s),\eta(s)),$$

and, moreover,

$$\pi(s) = \phi_x(\xi(s),\eta(s)),\rho(s) = \phi_y(\xi(s),\eta(s)).$$

To this end, consider the integral strip

$$B(s) = (\hat{\xi}(s),\hat{\eta}(s),\phi(\hat{\xi}(s),\hat{\eta}(s)),\phi_x(\hat{\xi}(s),\hat{\eta}(s)),\ \phi_y(\hat{\xi}(s),\hat{\eta}(s)))$$
$$\equiv (\hat{\xi}(s),\hat{\eta}(s),\hat{\zeta}(s),\hat{\pi}(s),\hat{\rho}(s)),$$

where $x = \hat{\xi}(s)$, $y = \hat{\eta}(s)$ is the unique solution through (x_o,y_o) of the first two equations of the characteristic system, when one substitutes $\phi(x,y),\phi_x(x,y),\phi_y(x,y)$ for u,p,q in F_p and in F_q.

One immediately verifies that $B(s)$ is a strip, in fact, an integral strip, and moreover $B(s) = E^*$. It is also evident that $B(s)$ lies in Σ. To prove the theorem, we show $B(s) = E(s)$. Actually, from (8), it follows that

$$\frac{d\hat{\zeta}}{ds} = \hat{\pi}(s)\cdot\hat{\xi}'(s) + \hat{\rho}(s)\cdot\hat{\eta}'(s) = F_p(s)\cdot\hat{\pi} + F_q(s)\cdot\hat{\rho}$$

$$\frac{d\hat{\pi}}{ds} = \frac{\partial\phi_x}{\partial x}\cdot\hat{\xi}'(s) + \frac{\partial\phi_x}{\partial y}\cdot\hat{\eta}'(s)$$

$$= \frac{\partial\phi_x}{\partial x}\cdot F_p(s) + \frac{\partial\phi_y}{\partial x}\cdot F_q(s)$$

$$= -F_x(s) - F_u(s)\cdot\hat{\pi}$$

$$\frac{d\hat{\rho}}{ds} = -F_y(s) - F_u(s)\cdot\hat{\rho}$$

Hence, $\hat{\xi},\hat{\eta},\hat{\zeta},\hat{\pi}$ and $\hat{\rho}$ satisfy the same differential system as ξ,η,ζ,π and ρ, and have the same initial values for $s = s_o$. Therefore, they are equal, q.e.d.

Observe that the proof is analagous to that of Theorem 3.1.

From Theorem 5.1, it immediately follows that the solution of the Cauchy problem is unique. Assuming now that the initial element E* depends on a parameter $r \in I_r$ in such a way that one has an initial strip such as $E^o(r)$ of the statement of the Cauchy problem in Article 1., then the following corollary is an immediate consequence of Theorem 5.1.

Corollary 5.1. If the Cauchy problem has a solution, it must be generated by juxtaposition of the characteristic strips $E(s;\ s_o,E^o(r))$ such that $E(s_o;\ s_o,E^o(r)) = E^o(r),\ r \in I_r$. Hence if there is a solution, it is unique.

In other words set

$E(s;\ s_o,E^o(r)) \equiv E(s,r) \equiv (\xi(s,r),\eta(s,r),\zeta(s,r),\pi(s,r),\rho(s,r))$.

If there exists an integral surface solution of the Cauchy problem, its parametric equations must be given by

$$x = \xi(s,r),\ y = \eta(s,r),\ u = \zeta(s,r)$$

$$(s,r) \in J = \{(s,r) \mid r \in I_r;\ -\delta_1(r) < s - s_o < \delta_2(r)\}$$

To obtain the integral surface in explicit form, it is

enough to solve the first 2 equations for s and r as functions of x and y and substitute in the third, getting

$$u = \zeta(s(x,y),\ r(x,y)) \equiv \phi(x,y)\ .$$

As we have proven in the remark, it is always possible to solve for s and r due to the transversality condition.

4. Solution of the Cauchy Problem.- Following is the statement of a theorem whose contents may be summarized: the Cauchy problem is a well posed problem. We now give the solution.

Theorem 5.2. Given the equation $F = 0$ and the initial integral strip $E^o(r)$, $r \in I_r$, respectively defined by (4,1) and (1-4), one has:

1) There exists an integral surface Σ,

$$\Sigma : u = \phi(x,y), (x,y) \in G, \tag{5,14}$$

through $E^o(r)$, that is, satisfying (5).

2) This surface is unique,

and 3) the surface depends continuously on the only initial data χ, σ and τ.

We omit the precise formulation and proof of 3), since it may be given in an analogous way to that of Theorem 3.2, although here, one must consider that $\sigma(r)$ and $\tau(r)$ are determined by application of the implicit function theorem to the system (3-4).

Part 2). has already been proven in Corollary 5.1 and the meaning of this uniqueness is the same as that already explained preceding the proof of Theorem 3.2.

Proof of 1). Let $E(s\ ;\ s_o, E^o(r)) \equiv E(s,r)$, for fixed r,

$r \in I_r$, be the characteristic strip which passes through $E^o(r)$ for $s = s_o$, that is to say, the unique solution of the characteristic system (10) given by

$$(5,15) \quad E(s;s_o,E^o(r)): \begin{cases} x = \xi(s;s_o,E^o(r)) \equiv \xi(s,r), \xi(s_o,r) = \alpha(r) \\ y = \eta(s;s_o,E^o(r)) \equiv \eta(s,r), \eta(s_o,r) = \beta(r) \\ u = \zeta(s;s_o,E^o(r)) \equiv \zeta(s,r), \zeta(s_o,r) = \chi(r) \\ p = \pi(s;s_o,E^o(r)) \equiv \pi(s,r), \pi(s_o,r) = \sigma(r) \\ q = \rho(s;s_o,E^o(r)) \equiv \rho(s,r), \rho(s_o,r) = \tau(r). \end{cases}$$

In these formulas, s and r vary over a domain J defined by (7a), and x and y vary over a domain G, the topological image of J by the homeomorphism H given in (7b). According to the remark, there is a domain G such that if we set $\zeta(s(x,y),r(x,y)) \equiv \phi(x,y)$, the surface Σ of (14) is well defined.

We prove now that the surface $u = \phi(x,y)$ thus defined is of class two. The characteristic system (10), when put in normal form, has second members which are of class one in D^5. Hence, by a well known theorem concerning the regularity of solutions of an ordinary differential system, it follows that the five functions of (15) are of class one with respect to all of their arguments. Solving the first two equations for s and r, and substituting the solutions in the other three, it follows, according to the regularity and existence theorem of implicit functions that

$$u,p,q \in C^1(G) .$$

But p and q are precisely the first derivatives of $u = \phi(x,y)$ and hence

$$\phi \in C^2(G),$$

q.e.d.

Finally, we must show that the surface $u = \phi(x,y)$ satisfies equation (4,1). To this end, it suffices to show that the integral elements $E(s,r)$, $(s,r) \in J$, either for fixed r and varying s, or for fixed s and varying r, constitute a strip, that is to say, it suffices to show respectively

$$(5,16a) \qquad M \equiv M_r(E(s)) \equiv \zeta_s - \pi\xi_s - \rho\eta_s = 0$$

$$(5,16b) \qquad N \equiv N_s(E(r)) \equiv \zeta_r - \pi\xi_r - \rho\eta_r = 0, \quad (s,r) \in J.$$

Now, the first of these two conditions is certainly satisfied, since for fixed r and varying s, the elements $E(s,r)$ generate a characteristic strip. It remains only to show (16,b).

To this end, considering the characteristic system (10), one has

$$(5,17) \qquad \begin{aligned} \frac{\partial N}{\partial s} &= \frac{\partial N}{\partial s} - \frac{\partial M}{\partial r} \\ &= \pi_r F_p(s,r) + \rho_r F_q + (F_x + \pi F_u)\xi_r + (F_y + \rho F_u)\,\eta_r . \end{aligned}$$

Now, the $E(s,r)$ are integral elements, and therefore $F(E(s,r)) = 0$. It follows,

$$(5,18) \qquad \frac{\partial F}{\partial r} = 0 \ , \ (s,r) \in J.$$

therefore,

$$(5,19) \qquad \frac{\partial N}{\partial s} = \frac{\partial N}{\partial s} - \frac{\partial M}{\partial r} - \frac{\partial F}{\partial r} = -F_u N.$$

Hence

$$N(s,r) = N(s_o,r)\cdot\exp\{-\int_{s_o}^{s} F_u(t,r)dt\} \quad ;$$

but $N(s_o,r) = 0$ for all $r \in I_r$, and therefore

$$N(s,r) = 0, \quad (s,r) \in J ,$$

q.e.d.

As an immediate consequence of the two preceding theorems we have the following corollary.

Corollary 5.2.- Let Σ_1 and Σ_2 be two integral surfaces (with all their tangent planes), let $B(E^o)$ be the characteristic strip through the element $E^o \subset \Sigma_1 \cap \Sigma_2$.

Then $B(E^o) \subset \Sigma_1 \cap \Sigma_2$.

Observations.- In stating the Cauchy problem, we assumed in (1) that $I_r = (a,b)$ is open and $|\alpha'| + |\beta'| > 0$. But, it is important to consider the following two special cases.

One may assume that Λ is a closed curve defined for $r \in \bar{I}_r = [a,b]$ such that $(\alpha(a),\beta(a),\chi(a)) = (\alpha(b),\beta(b),\chi(b))$, in order that Λ be continuous, and also that the first and second derivatives are continuous. Except for this, everything said is obviously true.

The second case is that in which the curve Λ reduces to a point M, that is, if $|\alpha'| + |\beta'| + |\chi'| = 0$, $r \in I_r$. In this case the second condition of (3) is automatically satisfied and the first condition says the initial integral strip reduces to the Monge cone or part of it. In this case, we can delete the point M so that the surface becomes an integral one, or we can generalize the concept of integral surfaces, but it is clear, except for this, all other results we obtained remain valid. In this case, the integral surface is called the *integral conoid* at M.

5. Geometric Interpretation of Characteristic Strips and Integral Surfaces.

Consider Figure 1. Note the supporting curve Λ and its projection Λ^* on the plane $u = u_1 - 1$, where $u_1 = u(M_1^*)$.

Let M_o be the supporting point of the initial integral element $E(s_o ; s_o, E^o(r_o)) = E(s_o, r_o)$, $r_o \in I_r$. For fixed r_o, consider the infinite family of Monge curves or strips originating at the element $E(s_o, r_o)$. Having constructed the element $M_o M_1 = ds$ corresponding to the element $E(s_o, r_o)$, the next differential element of the Monge curve must be a differential element of a generator of the Monge cone at the point M_1, which in figure 1 is represented by the cone M_1ABC.

We repeat the same process, but now taking as the origin the supporting point M_o' of the initial integral element $E(s_o ; s_o, E^o(r_o + dr)) = E(s_o, r_o + dr)$. Let $M_1' A'B'C'$ be the Monge cone at M_1'.

Now the only way that the two Monge strips, one starting at $E(s_o, r_o)$ and the other at $E(s_o, r_o + dr)$, will fit together and constitute a surface is that the strips chosen are those which are tangent through the Monge cones along the generators $M_1 B$ and $M_1' B'$ in such a way that $M_1 BB'M_1'$ be an element which is simultaneously tangent to both cones. Therefore M_oM_1B is the beginning of the characteristic curve originating at M_o and $M_o'M_1'B'$ is the beginning of the one orginating at M_o'. This is the desired interpretation.

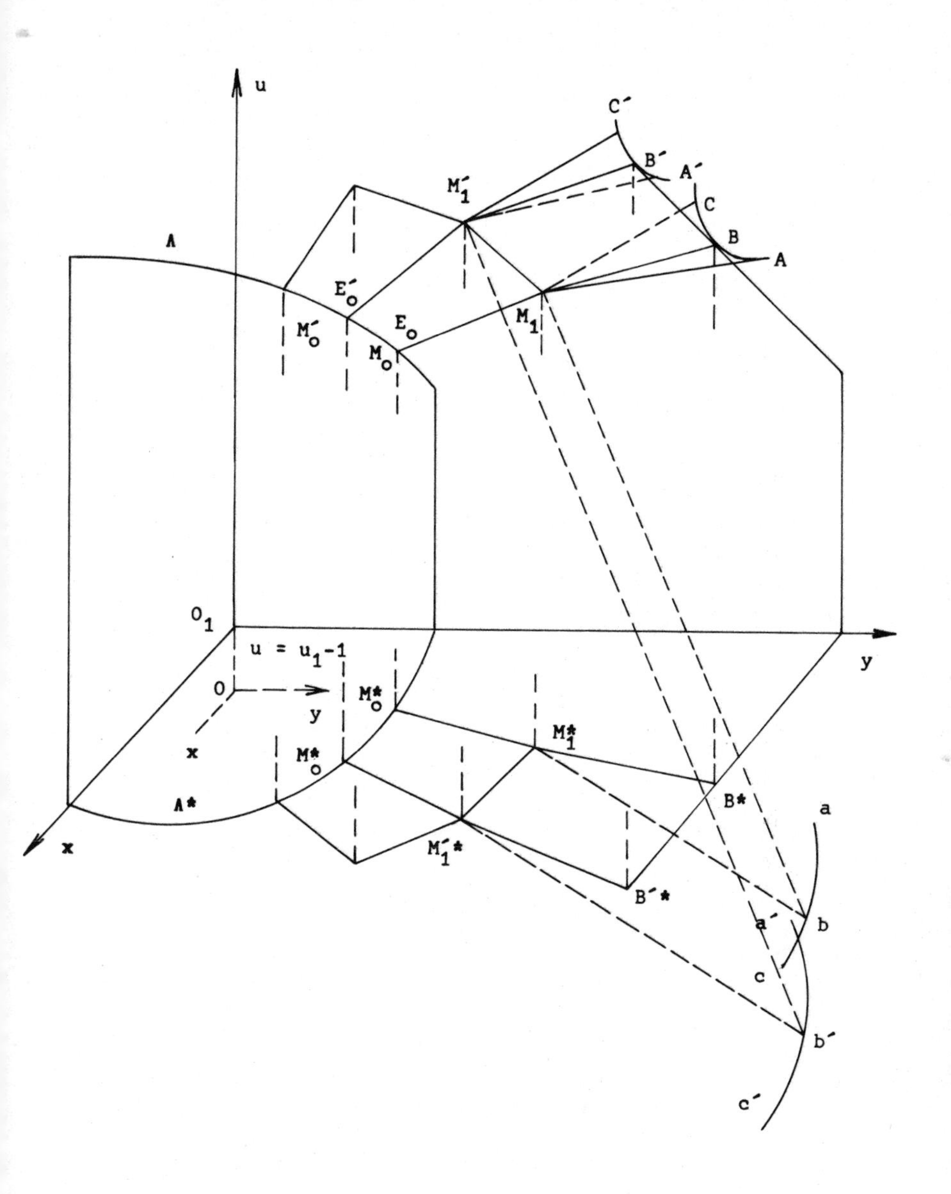

Figure 1

It follows that the characteristic strip originating at M_o is the intersection of the integral conoid at M_o with that at the point M_o' which is arbitarily close to M_o and on the initial curve Λ. That is to say, the characteristic strips through Λ on the intersection of integral conoids of points infinitely close to each other on Λ. Still in other words, the integral surface through Λ is the envelope of the integral conoids at the points of Λ.

This geometric interpretation offers a new way of considering the construction of an integral surface, through the given curve Λ, which is generated by Monge strips.

Actually, now the most obvious characterization of the characteristic strips is the stipulation

$$(5,20) \qquad \frac{\partial F(E(s,r))}{\partial r} = 0 \ , \ (s,r) \in J.$$

Of course, this condition, provided that the Monge strips according to (4,10) or (4,12) already automatically satisfy the condition

$$\frac{\partial F(E(s,r))}{\partial s} = 0, \ (s,r) \in J \ ,$$

is equivalent to (18). This determines the characteristic system and the characteristic strips. Conversely, if, in addition to condition (4,12), we stipulate condition (20) or (18), the Monge strips thus characterized generate the surface. Actually, condition (20) is equivalent to specifying the surface we want to construct as the envelope surface of the integral conoids of the points Λ, which satisfies all the required conditions; provided, of course, that one starts from a transverse initial strip, which is an integral one. If the initial strip $E^o(r)$ were such that $F(E^o(r)) = c \equiv$ constant, it is clear that the surface defined by equations (15) would be an integral surface of the equation $F = c$.

In Figure 1, we have shown the curve abc which, with respect to the axes $\{M_1*; p,q\}$ parallel to $\{O_1,x,y\}$, has the equation $F(M_1,p,q) = 0$.

Exercises

1. Verify that the theory developed does not apply, at least directly, to the equations

$$(p - q^2)^2 + (x - y)^2 + u^2 = 0, \quad (p + q^2)^2 + u^2 = 0,$$

$$(p - q^2)^2 = 0.$$

2. Suppose $F(x,y,u,p,q) = 0$ is not a quasi-linear equation. Then it cannot admit a general integral of the form

$$u = \sigma(x,y,f(\tau(x,y,u))),$$

where σ and τ are determined and f is arbitrary.

Conclude that the differential equation of first order (called the equation of tubular surfaces) which has as solutions the 2-parameter system of spheres

$$(x - a)^2 + (y - b)^2 + u^2 = 1,$$

cannot admit a general integral of the mentioned form.

3. Find the characteristic strips of the equation $p = q^2$. Find the integral surfaces through the initial curves

$$\begin{cases} x = 0 \\ u = 0 \end{cases}, \quad \begin{cases} y = x \\ u = 0 \end{cases}, \quad \begin{cases} y = 0 \\ u = x^2 \end{cases}.$$

4. Given the equation

$$p^2 + q^2 = 2(x^2 + y^2) + 4(x + y + 1) .$$

Let $C : x = \xi(s),\ y = \eta(s),\ u = \zeta(s)$, be a focal curve on the surface S,

$$S:\quad u = 1/2(x^2 - y^2) + xy + 2x ,$$

such that for $s = 0$, it passes through the point (x_o,y_o,u_o) where

$$u_o = 1/2(x_o^2 - y_o^2) + x_oy_o + 2x_o .$$

Do the following:

a) Determine the projection C* on the plane $\{0 ; x,y\}$ of the focal curve C.

b) Let $x_o = -1$, $y_o = 0$ and find a strip B which is a solution of the characteristic system passing through the element $E^o = (-1,0,-3/2,1,-1)$.

c) Indicate whether or not this strip is an integral one.

Remark: One may consider the more general case of the equation $p^2 + q^2 = g(x,y)$ and the surface $u = \varphi(x,y)$ where

$$\varphi_x^{\,2} + \varphi_y^{\,2} = (k^2 + 1) \cdot g(x,y) .$$

5. Suppose we are given the equation

$$p^2 + q^2 = f(x,y),$$

which admits as an integral surface $u = x^2 + y^2 + 7$. Consider the strip B which is a solution of the characteristic system passing through the element $E^o = (0,1,13,2,0)$.

a) Indicate whether or not the strip B is a characteristic strip.

b) Find the projection of the supporting curve of B on the plane $\{0\ ;\ x,y\}$.

6. Suppose we are given the equation

$$q^3 - H(x,y,p) = 0\ ,\ H(x,y,p) > 0,$$

and the initial curves

$$\Lambda_1 : x = r,\ y = 0,\ u = 1$$

$$\Lambda_2 : x = r,\ y = 0,\ u = 1 + r$$

$$\Lambda_3 : x = r,\ y = 0,\ u = 1 + r^2.$$

Let S_i be the integral surface through Λ_i, $i = 1,2,3$. Let B_i be the characteristic strip in S_i whose supporting curve passes through the point $(x = r = 0,\ y = 0,\ u = 1)$. Suppose also that

the strip B_1 passes through the element $E_1 = (0,5,7,p_1,q_1)$,

the strip B_2 passes through the element $E_2 = (0,5,7,p_2,q_2)$,

and the strip B_3 passes through the element $E_3 = (0,5,u_3,p_3,q_3)$.

Answer the following:

a) Is it necessary that $E_1 = E_3$?

b) Is it possible that $E_1 = E_2$?

c) If a characteristic strip passes through $E_o = (0,0,0,0,\sqrt[3]{H(0,0,0)}\ )$ and through $E^* = (0,5,u^*,p^*,q^*)$, must $E^* = (0,5,6,p_1,q_1)$?

7. Find the solution of $q^2 + px + y = 0$ passing through the curve Λ,

$$\Lambda : \begin{cases} x = r^2 \\ y = 1 \\ u = r \end{cases}.$$

Consider the characteristic curve C, lying on the above surface, through the point (1,1,1). Suppose $M = (\hat{x}, 3, \hat{u})$ is on the curve. Find $\hat{x}, \hat{u}$.

8. Find the integral surface of the equation

$$p^2 - q^2 - 2u = 0,$$

through the curve Λ,

$$\Lambda: \quad \begin{aligned} x &= 0 \\ u &= (1 + y)^2 \end{aligned} \quad .$$

What is the nature of the surface so obtained, that is, how can it be classified.

9. Find the general integrals in convenient domains of the equations

$$\frac{\partial^2 u}{\partial x^2} - \frac{\partial^2 u}{\partial y^2} = 0, \quad u \frac{\partial^2 u}{\partial x \partial y} - \frac{\partial u}{\partial x} \frac{\partial u}{\partial y} = 0 .$$

[Note that each of these equations, although of second order, consequently reduces to two successive first order equations. For the second, divide the equation by $u \cdot \frac{\partial u}{\partial x}$.]

10. Suppose $P = (0,0,c)$. Find the partial differential equation of first order of the surfaces such that the intersection of the tangent plane (at any point) with the perpendicular from P to the same plane is a point of the plane $\{0 ; x,y\}$. [Obtain a 2-parameter family of solutions, namely $x \cos \alpha + y \sin \alpha - cu/d = d$, and conclude that the equation is $u = xp + yq - c(p^2 + q^2)$.]

Prove that along a characteristic strip, p and q remain constant, and that the characteristic curves are straight lines.

Through elementary geometric considerations, it should be evident that the parabaloid of revolution $4cu = x^2 + y^2$ is a surface fulfilling the stated condition, and it actually is immediate to verify that it satisfies the equation. How can it be explained that this parabaloid is not generated by characteristic curves?

11. Given the equation $p^2 + q^2 = 1$, study the possibility of an integral surface containing the arc of the helix

$$x = \cos r \ , \ y = \sin r \ , \ u = r \ , \ 0 \leqq r \leqq \pi/2 .$$

[The helix is a Monge curve. The characteristic curves and strips are straight lines and planes forming an angle of $\pm \pi/4$ with the u-axis. Constructing the characteristic line through each point of the helix, one obtains two surfaces (the two sheets of the tangential developable helicoid) defined for $x,y \in \overline{G}$ where

$$G = \{(x,y) \mid 0 < x < 1, \ 0 < y < 1, \ 1 < x^2 + y^2\} .$$

Both surfaces are of the class $C^1(\overline{G}) \cap C^2(G)$, and both contain the initial curve. However, neither is a regular solution, or even a generalized one of the type of Ex. 7, § 3, since the initial curve is not interior to the closure of G, nor do there exist second derivatives of the surface along the initial curve.]

§6. The Method of the Complete Integral.

1. The Complete Integral.- In the preceeding § 5., we have seen how we can integrate the equation $F(x,y,u,p,q) = 0$ by the Cauchy method. We shall now study another method of solving the same equation, the method of Lagrange-Charpit, which offers some advantages, and is based on the concept of the complete integral, which of itself is interesting and adds much light to the sense and application of the equation $F = 0$.

In the theory of ordinary differential equations, one sees that the elimination of two independent parameters a, b in a 2-parameter family of curves

$$\phi(x,y,a,b) = 0$$

leads to a second order equation $F(x,y,y',y'') = 0$. Suppose now that there are two independent variables and consider the 2-parameter system of surfaces

$$(6,1) \qquad \phi(x,y,u,a,b) = 0 \ , \quad \frac{\partial\phi}{\partial u} \neq 0 \ ,$$

where ϕ is sufficiently smooth and where we consider u as a sufficiently smooth function of x and y. Suppose the two constants a and b are independent, one also says essential, that is to say, (1) cannot be put in the form

$$(6,2) \qquad \phi(x,y,u,a,b) = \phi(x,y,u,c(a,b)) \ ,$$

where in fact there is only one arbitrary constant c. For greater clarity, assume we have solved the equation for u, and consider

$$(6,3) \qquad u = \phi(x,y,a,b) \ .$$

The elimination of a,b through derivation leads to the following result. One has

$$(6,4) \qquad p = \phi_x(x,y,a,b) \ , \ q = \phi_y(x,y,a,b) \ .$$

In order that a,b can be eliminated from (3) and (4), it is sufficient that the matrix

$$(6,5) \qquad \begin{pmatrix} \phi_a & \phi_{xa} & \phi_{ya} \\ \phi_b & \phi_{xb} & \phi_{yb} \end{pmatrix}$$

be of rank two. Now, that this matrix be of rank two is precisely a necessary condition for a and b to be independent in the equation (3), that is, there does not exist $c(a,b)$ such that

$$\phi(x,y,a,b) = \psi(x,y,c(a,b)),$$

as is easily verified (Exercise 1.). Hence, a and b can be eliminated from (3) and (4), and the result clearly is an equation in u of the form

$$(6,6) \qquad F(x,y,u,p,q) = 0$$

in general, not linear or even quasi-linear.

It may be surprising that the elimination of two constants leads to a more complex equation than the elimination of an arbitrary function. For instance, if in the set of surfaces

$$(6,7) \qquad u = \phi(x,y,f(\psi(x,y))) \ ,$$

where ϕ and ψ are given functions, the arbitrary function f is eliminated, we arrive at a particularly simple semilinear equation

$$(6,8) \qquad \psi_y(x,y) \frac{\partial u}{\partial x} - \psi_x(x,y) \frac{\partial u}{\partial y} - h(x,y,u) = 0 \ ,$$

much simpler than (6). However, an arbitrary function is more indeterminate that two arbitrary constants. The paradox is explained considering that the function (7) is a "general integral" of equation (8), that is to say the solution of any Cauchy problem for (8) is contained in (7) and can be obtained for a particular function f. On the other hand, the function (3) is not a general integral of equation (6). The sense of function (3) is that whatsoever the numerical values of the parameters a and b be, one obtains a particular integral surface of equation (6), but, of course, this equation has many other integral surfaces which cannot be obtained from specific values of a and b. Nevertheless, Theorem 6.1 will show that the complete integral (3) contains implicitly, functional arbitrariness more complex than that of a simple explicit function $f(\psi)$.

Now suppose we are given the equation

$$(6,9) = (6,6) \qquad F(x,y,u,p,q) = 0,$$

where F satisfies the same conditions imposed at the beginning of ∮ 4.

Let

$$(6,10) = (6,3) \qquad u = \phi(x,y,a,b),$$

$$(x,y) \in G, \ (a,b) \in J, \ \phi \in C^2(K), \ K = G \times J,$$

where G and J are domains in the plane $\{0;\ x,y\}$, $\{0;\ a,b\}$ respectively, be a 2-parameter system of integral surfaces, that is, such that

$$(6,11) \qquad F(x,y,\ \phi(x,y,a,b),\ \phi_x,\ \phi_y) = 0, \ (x,y,a,b) \in K,$$

and such that the matrix (5) is of rank two for all $(x,y,a,b) \in K$; this insures the parameters a and b are independent. Then, by definition, one says the function $\phi(x,y,a,b)$ is a complete integral of the equation $F = 0$. Its introduction is due to Lagrange.

The importance of complete integrals lies in the following fact. Although it is clear that the complete integral (10) of (9) does not yield all regular integrals of (9) for some specific values of a and b, it does nevertheless yield all of these exclusively by differentiation and elimination techniques, without the necessity of solving any differential equation or carrying out quadratures. Geometrically speaking, the construction of this surface is carried out considering a 1-parameter system of integral surfaces of those contained in the complete integral, and constructing their envelope. The 1-parameter system is found by establishing a relation $b = \mu(a)$ between the parameters a and b. The possibility of getting all the (regular) integral surfaces without integration starting from a complete integral has its roots in the following fundamental property: if $\phi(x,y,u,a,b) = 0$ is a complete integral of $F = 0$ (defined in a domain D^3 of the arguments x,y,u), then it contains the whole 4-dimensional manifold of integral elements of $F = 0$ (supported by points of D^3); that is to say, given any integral element E^o, it is possible to specify $a = a_1$, $b = b_1$ in such a way that the integral surface $\phi(x,y,u,a_1,b_1)$ contains the integral element E^o, as follows immediately from the definition of complete integral.

Theorem 6.1. Let $u = \phi(x,y,a,b)$, *given by* (10) *be a complete integral of* (9) *and let*

$$b = \mu(a) \ , \ \mu \in C^2(I_a) \tag{6,12}$$

be an arbitrary relation between the parameters b *and* a, *where* a *lies in the interval* I_a; *then it follows:*

I. *The surface*

$$u = \phi(x,y) \ , \ (x,y) \in G_o \ , \tag{6,13}$$

generated by the curves

$$\begin{cases} u = \phi(x,y,a,\ \mu(a)) \\ o = \phi_a(x,y,a,\mu(a)) + \phi_b \cdot \mu'(a), (x,y) \in G_o \ , \end{cases} \tag{6,14}$$

when a *ranges over* I_a, *is a (real) integral surface of* $F = 0$ *provided there is a real point* (x_o,y_o,u_o) *which is a solution of* (14) *for* $a = a_o \in I_a$, *and such that*

$$\frac{\partial^2\phi(a_o)}{\partial a^2} \neq 0 \ . \tag{6,15}$$

II. *The system of integral strips depending on three arbitrary parameters* a,b,c *determined by*

$$\begin{cases} 0 = \phi_a + c \cdot \phi_b \\ u = \phi(x,y,a,b) \\ p = \phi_x(x,y,a,b) \\ q = \phi_y(x,y,a,b) \end{cases} \tag{6,16}$$

when the only independent variable x *or* y *is allowed to vary, satisfies the characteristic system of* $F = 0$ *and is therefore a 3-parameter system of characterisitic strips.*

III. *Given an initial strip*

$$E^{o}(r) = (\alpha(r), \beta(r), \kappa(r), \sigma(r), \tau(r)) \tag{6,17}$$

which satisfies conditions (5, 1-4) *of the Cauchy problem, one obtains the integral surface* $u = \psi(x,y)$, $(x,y) \in G_o$, *through* $E^{o}(r)$ *applying the first part of this theorem, taking as the relation of* b *and* a *that which results from eliminating* r *between*

$$\begin{cases} \kappa(r) = \phi(\alpha(r), \beta(r), a,b) \\ \sigma(r) = \phi_x(\alpha(r), \beta(r), a,b) \\ \tau(r) = \phi_y(\alpha(r), \beta(r), a,b) . \end{cases} \tag{6,18}$$

Remarks. It should be realized that all functions, whose existence is asserted by the theorem, exist in sufficiently small domains, since it is clear we deal here with local theorems. In particular, in (12) it is understood that for all $a \in I_a$ one has $(a,\mu(a)) \in J$, and analagously in other cases.

The final assumption of I is necessary since it might happen that the surface is imaginary, or even empty. See Ex. 6.

The second part should be understood thus: given an integral element $E^{o} = (x_o,y_o,u_o,p_o,q_o)$, there is a sufficiently small neighborhood of x_o or y_o, depending on which is the independent variable, in which the characteristic strip through

E^o is defined.

Therefore, in the previously mentioned 4-dimensional manifold of integral elements, which is immersed in the space $\{0;\ x,y,u,p,q\}$, the strips (16) contain all the characteristic strips of $F = 0$. This assertion is also an obvious consequence of III.

Considering the preceeding remarks and the results of ∮ 5, it is clear how part III must be taken. The system (18) is apparently overdetermined, but it must be compatible since we know already that there is such an integral surface.

Proof I. Having assumed the existence of a real point (x_o, y_o, u_o) which is a solution of (14) for $a_o \in I_a$, there exists a neighborhood of a_o for which the curves (14) have real intersection; and since we have

$$(6{,}19)\qquad \frac{\partial^2 \phi(a_o)}{\partial a^2} = [\phi_{aa} + 2\phi_{ab}\mu' + y_{bb}\mu'^2 + \phi_b \mu'']_{a = a_o} \neq 0 ,$$

the solutions (x,y,u) of (14) are such that (x,y) fill a neighborhood G_o of (x_o, y_o), and therefore there is a surface $u = \psi(x,y)$, $(x,y) \in G_o$.

This surface is an integral surface of $F = 0$. To prove it, we will show that all of its elements are integral, since each of them belongs to a characteristic strip. Let $a = a_1$ be a fixed value of a. The 1-parameter system of elements

$$(6, 20)\qquad \begin{cases} 0 = \phi_a(x,y,a_1,\mu(a_1)) + \phi_b(x,y,a_1,\mu(a_1)) \cdot \mu'(a_1) \\ u = \phi(x,y,a_1,\mu(a_1)) \\ p = \phi_x(x,y,a_1,\mu(a_1)) \\ q = \phi_y(x,y,a_1,\mu(a_1)) \end{cases}$$

is a characteristic strip. It is evident that they constitute a strip and that this strip is integral. Now, this integral strip lies in the evolute or integral surface $u = \phi(x,y,a_1,\mu(a_1))$, which is contained in the complete integral, and lies also in the envelope $u = \psi(x,y)$, since both surfaces are tangent along the curve (14) for $a = a_1$. Hence, by Corollary 5.2, it follows that this integral strip is a characteristic strip, q.e.d. One could also give an analytic proof which is left as Ex. 7.

II. The proof is almost contained in the given proof of I. Since, given the values a_1,b_1,c_1 of the parameters a,b,c, it is clear we can always choose μ so that $\mu(a_1) = b_1$, $\mu'(a_1) = c_1$, and hence system (20) becomes (16) for $a = a_1$, $b = b_1$, and $c = c_1$.

Given the integral element $E^o = (x_o,y_o,u_o,p_o,q_o)$, one may obtain the characteristic strip passing through E^o thus: substitute the given values of x,y,u,p,q into (16), then from the last three equations of (16) one may select two which can be solved for a and b as functions of E^o, since at least one of the second order minors of the matrix (5) must have a determinant, which is the Jacobian, different from zero. This gives the values of a and b. Call them a^* and b^*. Substituting these values in the first, one obtains the value of c, call it c^*. The strip defined by (16) when $a = a^*$, $b = b^*$, $c = c^*$ is the required strip.

Part III is a consequence of the first two since the integral surface is obtained by generating characteristic strips through the initial elements $E^o(r)$, $r \in I_r$. Hence, one must choose the parameters a,b,c in such a way that (16) is compatible with the equations $x = \alpha(r)$, $y = \beta(r)$, $u = \kappa(r)$, $p = \sigma(r)$, $q = \tau(r)$. Disregarding c, which appears only in the first equation of (16), we obtain precisely (18).

The system (18) is certainly compatible, that is, if one may eliminate r starting from two different pairs of equations, the relation $M(a,b) = 0$ to which one arrives will be the same in both cases. It is certainly not indeterminate, but yields a relation $M(a,b) = 0$ where $|M_a| + |M_b| > 0$. It may be that the system determines a and b as constants, namely when the given transverse initial strip lies in one of the integral surfaces contained in the complete integral. (See Ex. 4.). All this is a consequence of the Cauchy theorem proven in § 5. In particular, we leave as Ex. 8 to prove that if one obtains an equation independent of a and b, then it reduces to an identity.

2. Total Differential Equations.- Consider a relation of the form

$$(6,21) \qquad P(x,y)dx + Q(x,y)dy, \; P,Q \in C^1(D^2), \; |P| + |Q| > 0.$$

Let

$$(6,22) \qquad y = \phi(x,C), \; \frac{\partial\phi}{\partial C} \neq 0 ,$$

be a general integral of the equation

$$(6,23) \qquad Pdx + Qdy = 0 .$$

Solving (22) for C, one obtains

$$C = w(x,y) ,$$

where

$$\frac{w_x}{P} = \frac{w_y}{Q} = \mu(x,y) \ .$$

The system of functions $\mu \cdot M(w)$, M arbitrary, constitutes the set of <u>integrating factors</u> of (21), that is

$$M(w)\ (\mu P dx + \mu Q dy)$$

are exact differentials which when set equal to zero give equations equivalent to (23).

Consider now the total differential equation

(6,24) $\qquad du = P(x,y,u) \cdot dx + Q(x,y,u)\ du, P,Q \in C^1(D^3),$

assuming P and Q are known, and $u = \phi(x,y)$ is an unknown function, $\phi \in C^2(D^2)$, satisfying (24) for all $(x,y) \in D^2$.

To simplify notation, we represent partial derivatives through subindices as follows: P_x, Q_u represent partial derivatives assuming all arguments of P and Q vary independently, and we shall represent by $\frac{\partial P}{\partial x}$, $\frac{\partial Q}{\partial y}$ the composed partial derivatives, that is, assuming that only x and y are independent variables (the other variables of the function being functions of (x,y)).

Thus for instance

$$\frac{\partial P}{\partial x} = P_x + P_u \frac{\partial u}{\partial x} \equiv P_x + P_u \cdot u_x \ .$$

We say that (24) is <u>completely integrable</u> if

(6,25) $\qquad \frac{\partial P}{\partial y} = \frac{\partial Q}{\partial x}\ , \ (x,y,u) \in D^3 \ .$

From

$$du = \frac{\partial u}{\partial x}\, dx + \frac{\partial u}{\partial y}\, dy \ ,$$

and also by (24), we must have

$$(6,26) \qquad \frac{\partial u}{\partial y} = P(x,y,u) \ , \ \frac{\partial u}{\partial y} = Q(x,y,u), (x,y) \in D^2 \ .$$

Condition (25) may thus be written:

$$(6,27) \qquad P_y + Q \cdot P_u = Q_x + P \cdot Q_u, \ (x,y,u) \in D^3 \ ,$$

where all terms are known and the condition of complete integrability is easily decideable.

If (27) is satisfied, we shall see that there are solutions $u = \phi(x,y)$ of (24) depending on an arbitrary integration constant and that they may be obtained by integration of two ordinary differential equations of first order. However, it should be noted that it is not necessary that (27) be satisfied for integral surfaces of (24) to exist, since (27) may determine a surface $u = \psi(x,y)$ on which (27) is satisfied, but only for $(x,y) \in D^2$, and it may happen that this surface is a solution of (24); but in this case, the solution cannot depend on an arbitrary integration constant. See Ex. 12.

The property that an equation (24) be completely integrable is intrinsic in the following sense.

Assertion 6.1. *The condition of complete integrability is invariant relative to the changes of unknown functions.*

This says if the functions P and Q of (24) satisfy (27) and one substitutes $v = \psi(x,y)$ for $u = \phi(x,y)$ related by

$$(6,28) \qquad u = g(x,y,v) \ , \ g_v \neq 0 \ ,$$

one obtains the new total differential equation

$$(6,29) \qquad dv = P^*(x,y,v)\, dx + Q^*(x,y,v)\, dy ,$$

which is also completely integrable, that is

$$(6,30) \qquad P^*_y + Q^* \cdot P^*_v = Q^*_v + P^* \cdot Q^*_v , \quad (x,y,v) \in D^{3*} .$$

Proof. Carrying out the change of variables, we obtain a new equation in v of the form (29), where

$$(6,31) \quad \begin{cases} g_v \cdot P^*(x,y,u) \equiv P(x,y,g(x,y,v)) - g_x \\ g_v \cdot Q^*(x,y,u) \equiv Q(x,y,g(x,y,v)) - g_y . \end{cases}$$

Instead of verifying that (30) is fulfilled, which presents no special difficulty, it is shorter to take derivatives of the first of (31) with respect to y and of the second with respect to x. It follows immediately that (27), or equivalently (25), implies

$$\frac{\partial P^*}{\partial y} = \frac{\partial Q^*}{\partial x} .$$

Now this result is equivalent to (30) in the same fashion that (30) is equivalent to (27). The assertion is proven.

We proceed to obtain the general integral of (24) and prove simultaneously the existence of integral surfaces $u = \phi(x,y)$, $(x,y) \in G_o$, a domain in the plane $\{0;\ x,y\}$. Suppose that (24) is completely integrable. Then, if u exists, it must satisfy the first of (26), and hence, it follows

$$(6,32) \qquad u = \psi(x,y,k) , \ \psi_k \neq 0 ,$$

where $\psi(x,y,k)$ is the general integral of the ordinary differential equation

$$(6,33) \qquad \frac{du}{dx} = P(x,y,u) ,$$

where y is considered as a parameter. The integration constant k appearing as an argument of ψ, is constant in the sense it does not depend on x or u, but it may depend on y.

Substituting the u of (32) in equation (24), we have

(6,34) $$\psi_x dx + \psi_y dy + \psi_k\, dk = P\, dx + Q\, dy \ .$$

Since ψ is a solution of equation (33), we have $\psi_x = P$, and hence (34) reduces to

(6,35) $$\frac{dk}{dy} = \frac{Q(x,y,\psi(x,y,k)) - \psi_y(x,y,k)}{\psi_k(x,y,k)} \equiv \theta(x,y,k).$$

Now, if there is a solution, k may not depend on x as we have said. It implies that $\theta_x = 0$. Actually, (32) may be considered as a change of the unknown function, k being the new function. The new total differential equation obtained is (34) or (35) and can be written:

(6,36) $$dk = 0 \cdot dx + \theta(x,y,k)\, dy \ .$$

By virtue of Assertion 6.1, this equation is also completely integrable and therefore

$$\theta_x + \theta_k \frac{\partial k}{\partial x} = 0 \ .$$

But also according to (36) $\frac{\partial k}{\partial x} = 0$. Therefore, $\theta_x = 0$, q.e.d.

Therefore $\theta(x,y,k) \equiv \eta(y,k)$ and k is determined as the general integral of the ordinary differential equation

(6,37) $$\frac{dk}{dy} = \eta\ (y,k) \ .$$

Let $k = \chi(y,C)$ be the general integral depending on a real integration constant C which is independent of the variables x,y,u.

Hence, if there are solutions of (24), they are

included in the expression

(6,38) $\quad u = \psi(x,y, \quad \chi(y,C))$,

obtained by substituting k in (32).

It is immediate that $\frac{\partial\psi}{\partial x} = P(x,y,u)$ and that from (35), $\frac{\partial\psi}{\partial y} = Q(x,y,u)$. Therefore, (38) represents the general integral of (24).

These results are summarized in the following theorem

Theorem 6.2. The completely integrable total differential equation (24) admits a general integral given by (38), where $u = \psi(x,y,k)$ is the general integral of (33), in which y is considered as a parameter and k is the integration constant. In the same (38), $k = \chi(y,C)$ is the general solution of equation (37) in which $\eta(y,k)$ is the rational expression appearing in (35), which is independent of x. Finally, C is an integration constant.

It is easy to give a geometric interpretation of what equation (24) and the integral surfaces (28) represent in the 3-dimensional affine space referred to the Cartesian coordinate system $\{0; x,y,u\}$. The unknown u represents a surface Σ. Let (x_o,y_o,u_o) be a point M of Σ. Then (34) says the tangent plane through M to Σ is

(6,39) $\quad u - u_o \equiv (x-x_o)\cdot P(x_o,y_o,u_o)+(y-y_o)\cdot Q(x_o,y_o,u_o)$,

since the normal to the surface is given by the direction parameters $\left(\frac{\partial u}{\partial x} , \frac{\partial u}{\partial y} , -1\right)$.

The surfaces given by (38), integral surfaces of (24), are precisely those enjoying the property expressed by (39). This property depends on the normal at a generic point of the surface and is expressible in terms of the coordinates of that point. If equation (24), which represents this property, is not completely integrable, there will in general be no surface solution; if it is completely integrable, there will in general be a unique solution through a given point (x^*, y^*, u^*).

3. Finding a Complete Integral.- The characteristic system of the equation $F = 0$ is an ordinary differential system of fourth order which can be reduced to a third order system by eliminating one of the five variables via the relation $F = 0$. Therefore, to solve it in explicit and finite terms, we must first find a first integral of a third order system, and having this, the system can be reduced to a second order system depending on an integration constant appearing as a parameter.

Analogously, we find a first integral of the second order system with a constant, and finally, we integrate a first order equation containing two constants as parameters.

The final result contains three parameters which correspond to the three arbitrary parameters on which the system of characteristic strips depends.

We saw that knowledge of a complete integral permits us to determine the system of characteristic strips and to solve the Cauchy problem without integration.

We now give a method of finding a complete integral by which the Cauchy problem is solvable in a fashion possibly simpler than integrating the whole characteristic system.

Here, we find a first integral of the same third order system as before, that is, the characteristic system reduced or not by $F = 0$, but with the condition that the first integral

$w(x,y,u,p,q)$ satisfies

$$(6,40) \qquad \frac{\partial(F,w)}{\partial(p,q)} \neq 0 \ ,$$

that is, the Jacobian is non-zero.

Condition (40) implies the system

$$(6,41) \qquad F(x,y,u,p,q) = 0$$

$$(6,42) \qquad w(x,y,u,p,q) = a \equiv \text{constant}$$

can be solved for p and q yielding the two relations

$$(6,43) \qquad \begin{aligned} p &= \pi(x,y,u,a) \\ q &= \rho(x,y,u,a) \ . \end{aligned}$$

Note that whenever x,y,u,a lie in the domain of definition of the functions π and ρ, the element

$$(x,y,u,\ \pi(x,y,u,a),\ \rho(x,y,u,a))$$

is always an integral one of $F = 0$.

Relation (43) allows us to pose for the unknown function u the following total differential equation,

$$(6,44) \qquad du = \pi(x,y,u,a) \cdot dx + \rho(x,y,u,a) \cdot dy \ .$$

Now, by Theorems 5.2 and 6.1, we know there are integral surfaces of (44) depending on an arbitrary parameter b in addition to the parameter a. Such are the complete integrals of $F = 0$. Therefore, we might conjecture that (44) is a complete integral, and, that this is so is easily proven. It reduces to showing

$$\frac{\partial\pi}{\partial y} = \frac{\partial\rho}{\partial x}$$

that is,

$$\frac{\partial p}{\partial y} = \frac{\partial^2 u}{\partial x \partial y} = \frac{\partial q}{\partial x} \ .$$

The general integral of (44)

$$u = \phi(x,y,a,b)$$

is a 2-parameter system of integral surfaces and hence a complete integral.

In this method, we perform the same integrations as in the Cauchy method, but instead of finding a first integral of a 1-parameter system, we integrate an equation of the form

$$\frac{\partial u}{\partial x} = \pi(x,y,u,a)$$

depending on two parameters y,a.

There is also another method, of Mayer (see text of E. Goursat) which allows us to integrate the total differential equation (24) by integrating only one ordinary differential equation, but at the cost the equation depends on a second parameter.

In general, the recommended method is that of Cauchy due to its simplicity and often easier computations. It often gives a clearer and deeper insight into the geometric problems associated with the equation $F = 0$. On the other hand, for problems arising from the calculus of variations and the mechanics of holonomic systems with perfect constraints, it is preferable to use the method of the complete integral. For the study of hyperbolic systems, the Cauchy theory is essential.

4. Equation of the Monge Curves.- We now give another interesting application of the complete integral.

We propose to find a differential equation whose solution set is precisely the system of Monge curves of $F = 0$. It must be a relation between the coordinates x,y,u of a Monge curve and its derivatives, as functions of a variable s; one can take x if $F_p \neq 0$, or y if $F_q \neq 0$ as the independent variable. This equation has a solution set depending on an arbitrary function according to assertion 4.1. According to the same assertion, the equations of the Monge curves are

$$(6,45) = (4,10) \quad \begin{cases} F(x,y,u,p,q) = 0 \\ \dfrac{dx}{F_p((x,y,u,p,q)} = \dfrac{dy}{F_q} = \dfrac{du}{pF_p + q\,F_q} = ds. \end{cases}$$

From these three equations which must reduce to identities as (x,y,u) varies along a Monge curve, we eliminate p and q obtaining the desired equation. Let it be

$$(6,46a) \qquad M(x,y,u, \frac{dy}{dx}, \frac{du}{dx}) = 0,$$

if x is taken as the independent variable; if one wishes x,y,u be functions of s, we consider

$$\frac{dy}{dx}\,\frac{dx}{ds} = \frac{dy}{ds}, \qquad \frac{du}{dx}\,\frac{dx}{ds} = \frac{du}{ds}$$

and (6,46a) becomes

$$(6,46b) \qquad \hat{M}(x,y,u,\frac{dx}{ds}, \frac{dy}{ds}, \frac{du}{ds}) \equiv M(x,y,u,\frac{dy/ds}{dx/ds}, \frac{du/ds}{dx/ds}) = 0.$$

This ordinary differential equation in two unknown functions $M = 0$ is called the Monge equation, and is identically satisfied in x as (x,y,u) varies along a Monge curve. Given arbitrarily the projection $y = f(x)$ of a Monge curve, equation (46a) becomes

$$(6,47) \qquad M(x,f(x),u,f'(x), \frac{du}{dx}) = 0,$$

which is a first order ordinary differential equation.

Consider a fixed point (x_o, y_o, u_o) and a straight line

$$(6,48) \qquad \frac{x - x_o}{l} = \frac{y - y_o}{m} = \frac{u - u_o}{n}$$

through it. It is clear that

$$M(x_o, y_o, u_o, \frac{m}{l}, \frac{n}{l}) = 0$$

expresses a necessary and sufficient condition that the straight line (48) be a generator of the Monge cone at (x_o, y_o, u_o), that is to say

$$(6,49) \qquad M(x_o, y_o, u_o, \frac{y - y_o}{x - x_o}, \frac{u - u_o}{x - x_o}) = 0$$

represents in ordinary coordinates the Monge cone of vertex (x_o, y_o, u_o) the equation $F(x_o, y_o, u_o, p, q)$ represents in tangential coordinates. This duality establishes a correspondence between equations $M(x,y,u,y',u') = 0$ and $F(x,y,u,p,q) = 0$ and hence M enjoys the same arbitrarity as F. Passing from $M(x,y,u,y',u') = 0$ to the corresponding equation F we consider the direction parameters to the cone (49):

$$-M_{y'} \frac{y - y_o}{(x-x_o)^2} - M_{u'} \frac{u - u_o}{(x-x_o)^2}, \frac{M_{y'}}{x-x_o}, \frac{M_{u'}}{x-x_o},$$

or equivalently

$$\frac{M_{y'} \frac{dy}{dx} + M_{u'} \frac{du}{dx}}{M_{u'}}, \quad \frac{-M_{y'}}{M_{u'}}, \quad -1 .$$

Identifying these parameters with $(p,q,-1)$ we obtain two relations which together with $M = 0$ allow us to eliminate

y' and u' obtaining $F(x,y,u,p,q) = 0$.

The interesting part of this theory of the Monge equation, in connection with the complete integral of $F = 0$, is that given a complete integral, one may calculate the general solution of the Monge equation depending on an arbitrary function without recourse to integration.

Assertion 6.2 Let $u - \phi(x,y,a,b)$ be a complete integral of $F = 0$. Then the solution of the corresponding Monge equation (46a,b) or the equivalent system (4,10) is parametrically and implicitly given by the system of curves

$$(6,50) \quad \begin{cases} u = \phi(x,y,a,\mu(a)) \\ \dfrac{\partial\phi}{\partial a} \equiv \phi_a + \phi_b \mu' = 0 \\ \dfrac{\partial^2\phi}{\partial a^2} \equiv \phi_{aa} + 2\phi_{ab}\mu' + \phi_{bb}\mu'^2 + \phi_b\mu'' = 0 , \end{cases}$$

which depends on an arbitrary function $\mu(a)$ of the parameter a.

Proof. Actually, it is enough to consider that every Monge curve not a characteristic curve can be completed to a Monge strip, and one may consruct the semicharacteristics (i.e. for $s > s_o$ or $s < s_o$) which generate an integral surface bounded by the Monge curve which is the envelope of the characteristic curves lying in the integral surface. Equations (50) express precisely this property. See Ex. 16.

5. Singular Integrals.- Theorem 6.1 provides a method of solving the Cauchy problem for the equation

$$(6,51) \quad F(x,y,u,p,q) = 0,$$

via the construction of integral surfaces which are the

envelopes of 1-parameter systems of integral surfaces obtained for particular values of the complete integral

(6,52) $\quad u = \varphi(x,y,a,b), (x,y,a,b) \in K \subset E^4$,

through a relation $b = \mu(a)$.

It is clear we may consider the two parameters a,b of the complete integral as varying independently and we may expect there will be an envelope, and if it exists it is necessarily an integral surface. Of course, there must be something special about this integral surface, called a singular integral, since we have proven the existence and uniqueness of a solution to the Cauchy problem without considering it, and therefore also it cannot be a solution of the Cauchy problem, at least as we posed it in ∮ 5. On the other hand, all integral surfaces thus far considered are unique solutions of a Cauchy problem.

As an example, consider the <u>equation of the tubular surfaces</u>

$$u^2(p^2 + q^2 + 1) - 1 = 0 ,$$

which has as complete integral the set

$$u^2 = 1 - (x - a)^2 - (y - b)^2 \equiv \varphi^2(x,y,a,b)$$

of all spheres of radius 1 centered in the plane $\{0;\ x,y\}$. If we eliminate a and b between this equation and the relations $\varphi_a = \varphi_b = 0$, we obtain the surfaces

$$u = 1,\ u = -1,\ (x,y) \in E^2$$

which are evidently of class two and satisfy the differential equation, and therefore are integral surfaces. However, these surfaces may not have been considered in dealing with the Cauchy problem since here

$$F_p = F_q = 0 \text{ for } u = \pm 1 ,$$

and the condition $|F_p| + |F_q| > 0$ required for our treatment in (4,1) is not satisfied. Such integral surfaces are called singular integrals although in reality they have no singularities. Their elements are considered to be singular only because they do not satisfy condition (4,1).

The following assertion establishes the existence and uniqueness of singular integrals starting from the differential equation (51) and their connection with the complete integrals (52). We assume an equation $F = 0$ and domains D^5 and D^3, and a definition of integral surface with the same conditions as those of article 1, § 4. except for the assumption $|F_p| + |F|_q > 0$. We call <u>singular integrals</u> of (51) those integral surfaces

(6,53) $\quad u = \psi(x,y), \ (x,y) \in G ,$

such that every element is a <u>singular element,</u> that is to say,

(6,54) $\quad F_p(x,y,\psi(x,y),\psi_x,\psi_y) = F_q(x,y,\psi,\psi_x,\psi_y) = 0, \ (x,y) \in G.$

<u>Assertion 6.3. I)</u> <u>Given equation</u> (51), <u>let</u> $(x_o,y_o,u_o,p_o,q_o) \equiv E^o \subset D^5$ <u>be an element such that</u>

(6,55) $\quad F(E^o) = 0, \ F_u(E^o) \neq 0, \ F_{pp}(E^o)\cdot F_{qq}(E^o) \neq F^2_{pq}(E^o).$

<u>Then, there exists a unique integral surface</u>

(6,56) $\quad u = \psi(x,y), \ u_o = \psi(x_o,y_o), (x,y) \in G_o,$

where G_o is a neighborhood of (x_o,y_o), obtained by eliminating p and q between equations

$$(6,57)\qquad F = F_p = F_q = 0.$$

II) Given a complete integral (52) of equation (51), let $(x_o,y_o,a_o,b_o) \equiv P \subset K$ be such that

$$(6,58)\qquad u_o = \varphi(P),\ \varphi_{ax}(P)\cdot\varphi_{by}(P) \neq \varphi_{ay}(P)\cdot\varphi_{bx}(P),$$
$$\varphi_{aa}(P)\cdot\varphi_{bb}(P) \neq \varphi_{ab}^2(P)\ .$$

Then, there exists a unique singular integral (56) obtained by eliminating a and b between the equations

$$(6,59)\qquad u = \varphi(x,y,a,b),\ \varphi_a = \varphi_b = 0.$$

Proof I) The first condition of (55) expresses that there exists a 4-dimensional manifold of integral elements in a neighborhood of E^o in D^5, and insures the existence of at least one real element satisfying (57). The second condition of (55) implies one can solve F = 0 for u. The third condition insures the last two equations of (57) are solvable for p and q. Substituting these values in F = 0, we obtain the surface (56) which is evidently a singular integral.

Observe the following. First, the two conditions of (55) are also necessary since if $F_u(E^o) = 0$, no surface could be obtained in explicit form but only cylinders whose generators are parallel to the u - axis. Second, the singular integral is obtained from the given differential equation by differentiation and elimination techniques exclusively without recourse to integration, analogous to the case of ordinary differential equations.

<u>Proof II</u>) The three conditions of (58) are equivalent to the respective ones of (56). We leave as an exercise (see Ex. 1.) establishing the equivalence of the second of (58) with $F_u \neq 0$. The third condition of (58) insures the solvability of the last two equations of (59) for a and b; and substituting the values in the first, we get a surface such as (56). It remains only to prove such a surface is a singular integral.

Actually, since $u = \varphi(x,y,a,b)$ is a complete integral of $F = 0$, we have

$$F(x,y,\varphi(x,y,a,b),\ \varphi_x,\varphi_y) = 0,\ (x,y,a,b) \in K\ .$$

Taking derivatives with respect to a and b, we obtain

$$\begin{cases} F_u \cdot \varphi_a + F_p \cdot \varphi_{xa} + F_q \cdot \varphi_{ya} = 0 \\ F_u \cdot \varphi_b + F_p \cdot \varphi_{xb} + F_q \cdot \varphi_{yb} = 0\ . \end{cases}$$

Consider these two equations for the values of (x,y,a,b) that satisfy the last two equations of (59), that is, for the values corresponding to the surface

$$u = \psi(x,y),\ (x,y) \in G_o\ .$$

We obtain

$$\begin{aligned} &\varphi_{xa} \cdot F_p + \varphi_{ya} \cdot F_q = 0 \\ (6,60)\qquad &\varphi_{xb} \cdot F_p + \varphi_{yb} \cdot F_q = 0\ , \end{aligned}$$

which must be satisfied for all $(x,y) \in G_o$ and those values of a,b determined by $\varphi_a = \varphi_b = 0$, which are those corresponding to the surface $u = \psi(x,y)$. Now, consider (60) as a homogenous linear system whose coefficient determinant is non-zero by the second condition of (58). Hence, for the points of $u = \psi(x,y)$, $F_p = F_q = 0$, that is, (54) is satisfied and hence the surface is a singular integral , q.e.d.

The most important example of a singular integral is given by the equation of Clairaut of the form

$$u = xp + yq + J(p,q),$$

where J is given. Clairaut's equation admits the 2-parameter system of planes

$$u = ax + by + J(a,b)$$

as a complete integral. Refer to Ex. 17 for an exposition of the most important properties of this equation.

Exercises.

1. a) Prove that given

$$\varphi(x,y,a,b),\ \varphi \in C^2(D^4),$$

if there exists c(a,b) such that

$$(*)\ \varphi(x,y,a,b) = \psi(x,y,c(a,b)) ,$$

then the rank of

$$J = \begin{pmatrix} \varphi_a & \varphi_{xa} & \varphi_{ya} \\ \varphi_b & \varphi_{xb} & \varphi_{yb} \end{pmatrix}$$

is less than 2. And, conversely, if $r(J) < 2$, there is a function c(a,b) satisfying (*).

b) Prove

$$\begin{vmatrix} \varphi_{xa} & \varphi_{ya} \\ \varphi_{xb} & \varphi_{yb} \end{vmatrix} = 0$$

is a necessary and sufficient condition that $\varphi(x,y,a,b)$ admit an additive constant, that is, that there exist $a^*(a,b)$ and $b^*(a,b)$ with

$$\varphi(x,y,a,b) = \psi(x,y,a^*(a,b)) + b^*(a,b).$$

c) Prove $\varphi_{ya} = \varphi_{yb} = 0$ is a necessary and sufficient condition that there exist $a^*(a,b)$ and $b^*(a,b)$ satisfying

$$\varphi(x,y,a,b) = \psi(x,y) + \xi(x) \cdot a^*(a,b) + b^*(a,b) .$$

d) Let $u = \phi(x,y,a,b)$ be a complete integral of $F(x,y,u,p,q) = 0$. Prove $F_u = 0$ is a necessary and sufficient condition for $\varphi(x,y,a,b)$ to admit an additive constant.

2. Consider the set of surfaces

$$(*) \quad u = \varphi(x,y,u,f(\psi(x,y,u)))$$

where φ and ψ are fixed and f is arbitrary.

a) Prove that the differential equation of these surfaces obtained by eliminating f is quasi-linear. [From this comes part of the importance of the complete integral for non-semilinear equations].

b) Prove that for (*) to be of the form

$$\psi_y(x,y,u) \cdot p - \psi_x(x,y,u) \cdot q = 0$$

it is necessary and sufficient that the Jacobian matrix

$$\frac{\partial(\varphi,\psi)}{\partial(x,y,u)}$$

be of rank less than 2.

3. Find in finite terms the general equation of surfaces such that if through any point P one constructs the normal to the surface and it intersects the plane $\{0;\ x,y\}$ at the

point N, then $|\overrightarrow{ON}| = |\overrightarrow{NP}|$.

4. Eliminate the constants a,b of the 2-parameter system of surfaces

$$(x - a)^2 + (y - b^2) + z^2 = 1$$

and the 2 constants m and n of

$$(y - mx - u)^2 = (1 + m^2)(1 - z^2).$$

Verify one obtains the same differential equation [of tubular surfaces] and explain why.

Starting from the first complete integral, apply Theorem 6.1,III) and find the integral surfaces through the curve

$x^2 + y^2 = 1/4$, $u = \sqrt{3}/2$, and the element

$(1/2,\ 0,\ \frac{\sqrt{3}}{2},\ \frac{-\sqrt{3}}{3},\ 0)$

and, through the curve

$y = u = \sqrt{2}/2$, and the element $(0,\ \frac{\sqrt{2}}{2},\ \frac{\sqrt{2}}{2},\ 0,\ -1)$.

Finally, do the same with the second complete integral.

5. Given

$$pq = 4xyu,$$

verify that

$$u = (x^2 + a)(y^2 + b)$$

is a complete integral. Find the envelope of the 1-parameter system of surfaces obtained setting a = b, and verify that it is also a solution.

6. Consider the 2-parameter system of spheres

$$u^2 = a^2 - x^2 - (y - b)^2 \equiv \varphi^2(x,y,a,b).$$

Eliminating the parameters we obtain

$$u\,p + x = 0.$$

By the Cauchy method, we obtain the characteristic curves as the circumferences

$$y = y_o, \; x^2 + u^2 = x_o^2 + u_o^1 .$$

Applying (16), one obtains the system of characteristic strips

$$\begin{cases} c(y - b) - a = 0 \\ u^2 + x^2 + (y - b)^2 - a^2 = 0 \\ u\,p + x - 0 \\ u\,q + (y - b) = 0. \end{cases}$$

The first two yield

$$y = \frac{a}{c} + b \equiv a^*, \; x^2 + u^2 = a^2 - \frac{a^2}{c^2} = b^* .$$

Find the 4 integral surfaces corresponding to the relations

$$a^2 = 2b^2, \; a^2 - b^2 = 1, \; a^2 \doteq b^2 - 1.$$

[The "surfaces" are respectively: an imaginary cone, and two circumferences, the first real, the second imaginary and both in the plane $y = 0$ (the radical plane of the spheres). Verify that the conditions of theorem 6.1 are not satisfied. Observe the equation and much of the rest would be the same if we had written

$$u^2 = a^2 - x^2 + f(y,b).]$$

7. Carry out the analytic proof indicated at the end of the proof of the first part of Theorem 6.1. [One may see it generalized in Theorem 9.1.]

8. Carry out the proof indicated at the end of article 1.

9. From an equation $F(x,y,u,p,q) = 0$ suppose we know $u = g(x,y,a)$ is a 1-parameter system of solutions which admits the surface

$$u + \frac{(1 + y)^2}{4x} = 0$$

as envelope. In addition suppose one knows $g(x,y,1) = x + y + 1$.

Is it possible to prove there is a characteristic strip whose supporting curve passes through $(1,-3,-1)$?

10. Suppose we are given

$$yu\,dx + xu\,dy + f(x,y)du = 0.$$

Find the most general form of f such that:

a) the first member is an exact differential,

b) the equation is completely integrable.

c) Integrate when $f(x \cdot y) = x^2y^2 + xy$.

11. Let $V = \{f(x,y,u),\ g(x,y,u)\ h(x,y,u)\}$ be a vector field where f,g,h are the components with respect to cartesian coordinates.

Prove that for a surface $u = u(x,y)$ to be orthogonal to the field curves it is necessary and sufficient that

$$(*)\ f dx + g dy + h du = 0.$$

[That is to say, the vector be orthogonal to the arc element on the surface.]

Prove that the condition of complete integrability of (*) is

$$\frac{\partial}{\partial y}\left(-\frac{f}{h}\right) = \frac{\partial}{\partial x}\left(-\frac{g}{h}\right),$$

[which implies the elementary rectangle dx,dy closes].

It may also be expressed by requiring the existence of a function (potential) $\Phi(x,y,u)$ and of a function $a(x,y,u)$ such that

$$\text{grad } \Phi = \{af,\ ag, ah\}.$$

12. Find the <u>orthogonal surfaces</u> to the field curves of the vector field

$$W = \{u^2, x^3y,\ -\ x^2y\}.$$

[The only such surface is $u = xy$.]

13. In (35), k is independent of x and u. Taking derivatives with respect to x, prove $\theta_x = 0$.

14. Find the complete integral of the equation

$$p = (qy + u)^2.$$

15. Find a complete integral of $p = q^2$, and the integral surfaces through the curve

$$x = 1,\ y = t,\ u = t^2\ .$$

16. Suppose <u>the equation of light rays</u> in a homogenous 2-dimensional continuum is

$$p^2 + q^2 = \nu^2\ .$$

Prove that the Monge equation of the focal curves is

$$du^2 - \nu^2(dx^2 + dy^2) = 0.$$

Obtain a complete integral $[u = \nu x \cos a + \nu y \sin a + b]$ and verify that

$$\nu^2 x^2 = \mu'(a) \cdot \sin a + \mu''(a) \cdot \cos a$$

$$\nu^2 y^2 = -\mu' \cdot \cos a + \mu'' \cdot \sin a$$

$$u = \mu(a) + \mu''(a),$$

are the equations of the focal curves.[With $\nu = 1$ and reduced coordinates, they are the equations of the curves of zero length in the Minkowski space].

17. Consider the Clairaut equation

$$u = x\,p + y\,q - c(p^2 + q^2)$$

where c is a fixed positive number.

Prove that the characteristic curves and strips are straight lines and planes respectively. Hence, all regular integral surfaces are developable. Find the singular integral [see Ex. 10, § 5.]. Prove that the characteristic straight lines are all and only those tangent to the singular integral, and that the regular integral surfaces are all and only those developable surfaces tangent to the singular integral along an arbitrary curve. Find a simple characterization for solvability of the Cauchy problem.

§7. Example: $p^2 + q^2 + f(x,y) = 0.$

Consider the equation

$$(7,1) \qquad F(x,y,u,p,q) \equiv p^2 + q^2 - \left(\frac{x^2}{h^2} + \frac{y^2}{k^2}\right) = 0,$$

where later on we will assume $h = k$.

We consider:

1. Regularity of the equation and its motivation.

2. The characteristic system and its solution. There is only a 2-parameter system of base characteristic curves.

3. Finding a complete integral, and from it, the characteristic curves.

4. A transverse initial strip with $u = 0$. Transversality condition. Orthogonality of the initial curve to the characteristics. Initial strips, transverse or singular.

5. Integral surfaces. Integral conoids at $(x_o,y_o,u_o) = (0,y_o,0)$. Level curves or wave fronts, their orthogonality with the characteristics. Integral surfaces defined by other initial strips.

6. Connection with Fermat's principle and calculus of variations. Parallelism of the wave fronts and Huyghens' construction.

1. Equation (1) is closely related to the wave equation

$$(7,2) \qquad \frac{\partial^2 v}{\partial x^2} + \frac{\partial^2 v}{\partial y^2} - \left(\frac{x^2}{h^2} + \frac{y^2}{k^2}\right)\frac{\partial^2 v}{\partial t^2} = 0.$$

Of the numerous images or physical phenomena for which (2) is a model, we consider two.

The cartesian coordinates (x,y) are those of the projection of a point of an elastic membrane (soap bubble or a stretched elastic band), almost horizontal, and $v = v(x,y,t)$ is the form of the membrane as time varies, that is, the motion or vibration of the membrane. We assume the membrane has constant tension τ at all points and a surface density ρ, which varies with the point, whose value is

$$\rho = \tau\left(\frac{x^2}{h^2} + \frac{y^2}{k^2} \right) .$$

The coordinates (x,y) may also be interpreted as those of a point in a plane crystal referred to a cartesian system $\{O;\ x,y\}$. Then v is the intensity of a monochromatic light ray at this point. The crystal has an index of refraction given by

$$\left(\frac{x^2}{h^2} + \frac{y^2}{k^2} \right)$$

at a point (x,y).

The (base) characteristics of (1) are the (paths) trajectories of the rays along which energy and motion is propagated. The characteristic surfaces of (1) give the successive wave fronts of the motion governed by (2). The velocity of the wave propogation measured along the rays is

$$\left(\frac{x^2}{h^2} + \frac{y^2}{k^2} \right)^{-1/2} .$$

In both physical phenomena, the unknown u of (1) is precisely the time t of equation (2).

We will consider the membranes, since the physical realization of the crystal is more limited. In principle, the membrane can be extended to a sufficiently large portion of the plane $\{0;\ x,y\}$ minus the point $(0,0)$ where the density is zero. Therefore, the domain D^2 is the plane punctured at the origin,

$$D^2 = \{(x,y) \mid x^2 + y^2 > 0\} .$$

Examining (1), we see that F is analytic in all its arguments and the regularity condition

$$|F_p| + |F_q| > 0$$

reduces to

$$|p| + |q| > 0,$$

which is equivalent to $x^2 + y^2 > 0$, that is, we again find that the domain of (x,y) is the plane punctured at $(0,0)$ and u,p,q may vary between $-\infty$ and $+\infty$.

2. The characteristic system of (1) is

$$(7,3) \qquad \frac{dx}{p} = \frac{dy}{q} = \frac{du}{\frac{x^2}{h^2} + \frac{y^2}{k^2}} = \frac{dp}{\frac{x}{h^2}} = \frac{dq}{\frac{y}{k^2}} = ds .$$

This system admits a general integral depending on four arbitrary parameters. Assuming equation (1), we know the characteristic strips depend on three arbitrary parameters. For example, fixing $x = x_o$ and taking s an auxiliary parameter, we can find the unique solution such that for $x = x_o$, we have $y = y_o$, $u = u_o$, $p = p_o$ where y_o, u_o, p_o are arbitrary and q is determined by

$$q = \pm \left(\frac{x_o^2}{h^2} + \frac{y_o^2}{k^2} - p_o^2 \right)^{1/2} \equiv \pm q_o .$$

Equation (1) is not directly dependent on u, that is, $F_u = 0$. Hence, if $u = \varphi(x,y)$ is an integral surface, so is $u = u_o + \varphi(x,y)$. The relation $F(x_o,y_o,u_o,p_o,q_o) = 0$ of the components of the integral elements is independent of u_o ; since moreover, the characteristic system is autonomous with respect to u, the general integral can be put in a form such that u_o always appears in an expression $(u - u_o)$. The same is true for $(s - s_o)$ if one takes s as the independent variable. It follows that the characteristic strips are invariant with respect to translations parallel to the u-axis, and hence there is only a 2-parameter system of base characteristics.

In system (3), we also have

$$du = \left(\frac{x^2}{h^2} + \frac{y^2}{k^2} \right) ds$$

due to the fact that

$$pF_p + qF_q = 2(p^2 + q^2) = 2\left(\frac{x^2}{h^2} + \frac{y^2}{k^2} \right)$$

by virtue of $F = 0$. Now, by Euler's theorem on homogeneous functions, a sufficient condition that through $F = 0$ we can eliminate p and q in $\frac{du}{ds}$ is

$$F(x,y,u,p,q) = F_1(x,y,u,p,q) + F_2(x,y,u)$$

where F_1 is homogeneous in p and q.

If in (3) we do not consider the equation in du and we take s as the independent variable, there are 4 equations forming a homogeneous linear differential system with constant coefficients

$$(7,4) \qquad \frac{d}{ds}\begin{pmatrix} x \\ y \\ p \\ q \end{pmatrix} = \begin{pmatrix} 0 & 0 & 1 & 0 \\ 0 & 0 & 0 & 1 \\ h^{-2} & 0 & 0 & 0 \\ 0 & k^{-2} & 0 & 0 \end{pmatrix} \begin{pmatrix} x \\ y \\ p \\ q \end{pmatrix},$$

whose integration offers no difficulty.

We find the two first integrals

$$\frac{1}{h^2} x^2 - p^2 = C_1 , \quad \frac{1}{k^2} y^2 - q^2 = C_2$$

and imposing $F = 0$, we find $C_1 = -C_2 = C$,

$$(7,5) \qquad p^2 = \frac{x^2}{h^2} - C , \quad q^2 = \frac{y^2}{k^2} + C.$$

The relation between the integration constant C and the components of the initial element E^o is

$$(7,6) \qquad C = \frac{x_o^2}{h^2} - p_o^2 = - \left(\frac{y_o^2}{k^2} - q_o^2 \right).$$

Eliminating p and q in the first two equations of (3), we find

$$(7,7) \qquad \frac{dx}{\varepsilon\sqrt{\frac{x^2}{h^2} - c}} = \frac{dy}{\delta\sqrt{\frac{y^2}{k^2} + c}} = ds,$$

$$\varepsilon = \pm 1, \quad \delta = \pm 1 .$$

Integrating as functions of s and introducing initial values, we get

$$(7,8)\quad \begin{cases} s - s_o = h \log \dfrac{x + \varepsilon\sqrt{x^2 - Ch^2}}{x_o + \varepsilon\sqrt{x_o^2 - Ch^2}} \\ s - s_o = k \log \dfrac{y + \delta\sqrt{y^2 + Ck^2}}{y_o + \delta\sqrt{y_o^2 + Ck^2}} \end{cases}$$

hence

$$(7,9)\quad \begin{cases} x = x_o \cos \dfrac{s-s_o}{h} + \varepsilon\, h\, p_o \sin \dfrac{s-s_o}{h} \\ \quad \equiv \xi(s;s_o,x_o,y_o,u_o,p_o,q_o) \\ y = y_o \cos \dfrac{s-s_o}{k} + \delta\, k\, q_o \sin \dfrac{s-s_o}{k} \\ \quad \equiv \eta(s;s_o,x_o,y_o,u_o,p_o,q_o), \\ \qquad \varepsilon = \delta = +1, \\ p_o^2 + q_o^2 = \dfrac{x_o^2}{h^2} + \dfrac{y_o^2}{k^2}\ . \end{cases}$$

Taking derivatives of both with respect to s, and setting $s = s_o$, it follows that $\varepsilon = \delta = +1$. Furthermore, we assume $h, k > 0$.

Equations (9), (5), (6) constitute the general integral of (4) as functions of initial values.

To obtain the base characteristics given by (9), and of course it can be verified they depend on two parameters, in terms of an equation $\theta(x,y) = 0$, we eliminate $(s-s_o)$ in (8) obtaining algebraic curves when h/k is rational. If $h = 2k$, θ is a polynomial of second degree in y, but even so, the functions are difficult to handle.

Now, set $h = k$. We obtain the implicit equation $\theta(x,y) = 0$ of the base characteristics depending on two arbitrary constants. We can assume x is the independent variable and set $x_o = 0$. (With this, if we consider only real values, the only solutions lost are those not cutting the y-axis). Moreover, if we set

$$(7,10) \quad \begin{cases} p_o = \left(\frac{x_o^2 + y_o^2}{k^2}\right)^{1/2} \cdot \cos r = \frac{y_o}{k} \cos r \\ q_o = \frac{y_o}{k} \sin r \, , \end{cases}$$

then (9) takes the form

$$(7,11a) \quad \begin{cases} x = y_o \cos r \sinh \frac{s-s_o}{k} \\ y = y_o \cosh \frac{s-s_o}{k} \sin r \sinh \frac{s-s_o}{k} \quad . \end{cases}$$

It follows that

$$x^2 + y^2 + y_o^2 = 2y_o y \cosh \frac{s-s_o}{k} \quad ;$$

combining this with the first equation of (11), we can eliminate $s - s_o$, obtaining

$$(7,11b) \quad (x^2 + y^2 + y_o^2)^2 - \frac{4x^2y^2}{\cos^2 r} = 4y_o^2y^2 \, ,$$

that is

$$(x^2 - y^2 + y_o^2)^2 = 4 \, tg^2 r \cdot x^2y^2 \, ,$$

and finally

$$(7,12) \quad y^2 - x^2 - 2 \, tg \, r \cdot xy - y_o^2 = 0 \, .$$

Hence, the base characteristics constitute a 2-parameter system of equilateral hyperbolas centered at the origin and cutting the y axis precisely at $(0,y_o)$. Now, setting $y_o^2 = \overline{y}_o$, $-\infty < \overline{y}_o < \infty$, we obtain all the base characteristic curves. Since $\overline{y}_o$ is arbitrary, we can make a change of variables such that all these hyperbolas pass through (x_o,y_o). Moreover, if by (10) we set

$$\text{(7,13)} \qquad \operatorname{tg} r = \frac{q_o}{p_o} = C_o ,$$

from (12) it follows

$$\text{(7,14)} \qquad \theta(x,y;\ x_o,y_o,C_o) \equiv y^2-x^2-2C_oxy - (y_o^2-x_o^2-2C_ox_oy_o) = 0.$$

If one desires that the system of base characteristics be functions of $x_o,y_o,y_o' \equiv \dfrac{dy(x_o)}{dx}$, it easily follows that

$$\text{(7,15)} \qquad \begin{aligned} \theta_o(x,y;\ x_oy_o,y_o') &\equiv (x_oy_o' + y_o)(y^2 - x^2) - \\ &- 2(y_oy_o' - x_o)xy - (x_o^2 + y_o^2)(x_oy_o' - y_o) = 0. \end{aligned}$$

Taking derivatives of (14) twice with respect to x and eliminating C_o, we obtain the differential equation

$$\text{(7,16)} \qquad (x^2 + y^2)y'' + xy'^3 - yy'^2 + xy' - y = 0$$

of the base characteristics.

The values of p,q corresponding to a point (x,y) of the base characteristic are given by (5-6). It remains to determine the value of u. Suppose h = k. The characteristic system (3), (9) with (6) give

$$du = \int_{s_o}^{s} \frac{x^2 + y^2}{k^2} \, ds$$

$$(7,17) \qquad = \frac{1}{k^2} \int_{s_o}^{s} [(x_o^2 + y_o^2) \cosh \frac{2(s-s_o)}{k} + k(x_o p_o + y_o q_o) \sinh \frac{2(s-s_o)}{k}] \, ds.$$

Integrating, we obtain

$$u = u_o + \sinh \frac{s-s_o}{k} \left(\frac{x_o^2 + y_o^2}{k} \cosh \frac{s-s_o}{k} + (x_o p_o + y_o q_o) \sinh \frac{s-s_o}{k} \right) \qquad (7,18)$$

$$\equiv \zeta(s;\, s_o, x_o, y_o, u_o, p_o, q_o).$$

Equations (9), (18) and (5-6) constitute the general integral of the characteristic system (3); they give explicitly the characteristic strip of (1) which for $s = s_o$ passes through the element $E^o = (x_o, y_o, u_o, p_o, q_o)$, $k^2(p_o^2 + q_o^2) = x_o^2 + y_o^2$.

There is no difficulty in expressing u as a function of x,y and E^o. By (9), we can eliminate $(s-s_o)$ in (18) obtaining

$$(7,19) \qquad u = u_o + \frac{(y_o x - x_o y)(x_o x - y_o y)}{k^2 (x_o q_o - y_o p_o)}$$

and of course one could also obtain u as a function of x only applying (15).

To conclude the study of characteristic strips we make the

following remark. At each point (x_o, y_o, u_o), $|x_o|+|y_o| \neq 0$, the Monge cone is the right cone of revolution whose axis is parallel to the u-axis

$$(7,20a) \qquad p^2 + q^2 = \frac{x_o^2}{h^2} + \frac{y_o^2}{k^2},$$

and whose generators are by (4,7)

$$(7,20b) \qquad \frac{x - x_o}{p} = \frac{y - y_o}{q} = \frac{u - u_o}{\frac{x_o^2}{h^2} + \frac{y_o^2}{k^2}} .$$

Calling ω the angle these generators make with the u-axis of time, we obtain

$$(7,21) \qquad \tan \omega = \tan \arccos \frac{\frac{x_o^2}{h^2} + \frac{y_o^2}{k^2}}{\sqrt{\frac{x_o^2}{h^2} + \frac{y_o^2}{k^2} + \left(\frac{x_o^2}{h^2} + \frac{y_o^2}{k^2}\right)^2}} = \left(\frac{x_o^2}{h^2} + \frac{y_o^2}{k^2}\right)^{-1/2}$$

The quantity $\tan \omega$ has the dimension of velocity, and at each point represents the velocity with which the energy, the motion, the wave front propogates along a characteristic curve.

It is clear that the tangent plane to the Monge cone at (x_o, y_o, u_o) along any generator forms the same angle ω with the u-axis. Hence the Monge curves and strips are characterized by the following property. A Monge curve is that one for which the tangent straight line at each of its points (x,y,u) forms an angle with the u-axis whose tangent is given by

$$(7,22)\qquad \frac{ds^*}{du} = \tan\omega = \left(\frac{x^2}{h^2} + \frac{y^2}{k^2}\right)^{-1/2}, \quad (ds^*)^2 = dx^2 + dy^2.$$

A Monge strip is that supported by a Monge curve such that at each point (x,y,u), the associated plane forms the same angle ω with the u-axis that the straight line tangent to the supporting curve forms. It is therefore clear how, given any curve in the plane $\{0;\ x,y\}$, one can construct the only two Monge strips (symmetric with respect to the plane $u = 0$, and up to a constant) that have as a base Monge curve the given curve.

Of course, the property of forming the same angle ω with the u-axis is also enjoyed by the characteristic curves and strips, which are particular cases of Monge curves and strips.

The base characteristic curves are the paths traced out by the rays, that is, the visible traces in the crystal or membrane. The characteristic curves in the space $\{0;\ x,y,u\}$ are the trajectories of the same rays which define the motion, that is, their position as a function of time.

3. Since the variables in equation (1) can be separated, obtaining a complete integral is easy. Equation (1) may be written

$$(7,23)\qquad p^2 - \frac{x^2}{h^2} = \frac{y^2}{k^2} - q^2.$$

Setting

$$u = \overline{\phi}(x) + \overline{\psi}(y)$$

and substituting in (23), we get

$$\overline{\Phi}'^2 - \frac{x^2}{h^2} = -\overline{\Psi}'^2 + \frac{y^2}{k^2} = -a \equiv \text{constant}.$$

Therefore, we must have

$$(7,24) \qquad p = \frac{d\overline{\Phi}}{dx} = \sqrt{\frac{x^2}{h^2} - a}\ , \quad q = \frac{d\overline{\Psi}}{dy} = \sqrt{\frac{y^2}{k^2} + a}\ ,$$

hence the complete integral

$$(7,25a) \qquad u = \phi(x,y,a,b) \equiv \overline{\Phi}(x,h,a) + \overline{\Phi}(y,k,-a) + b,$$

where

$$(7,25b) \qquad \overline{\Phi}(x,h,a) = \frac{x}{2h}\sqrt{x^2 - ah^2} - \frac{ha}{2}\log(x + \sqrt{x^2 - ah^2}).$$

The system of characteristic curves depending on three arbitrary parameters a,b,c, is given by (25) and

$$(7,26) \qquad \overline{\Phi}_a(x,h,a) - \overline{\Phi}_a(y,k,-a) + c = 0.$$

These curves are completed to characteristic strips by the addition of (24) to (25-26). Equations (24) are the same as (5) and, as these, must be completed with the addition of (6).

Given an initial integral element, the characteristic strip through it is determined by the values of a,b,c. Equation (24) gives the value of a,

$$(7,27) \qquad a = \frac{x_o^2}{h^2} - p_o^2 = q_o^2 - \frac{y_o^2}{k^2}\ .$$

Now, given a, by (25)

$$(7,28) \qquad u_o = \phi(x_o,y_o,a,b)$$

and b is determined; c is determined by (26).

The 2-parameter system of base characteristics is directly given by (26). It is left as exercise 3. to verify that this system is equivalent to the one given by (12) or (14).

Having determined the base characteristics, their equations together with (25) determine the characteristic curves, which are completed to characteristic strips with (24).

4. Initial Strips. a). We look for a transverse initial strip $E^o(r)$, whose supporting curve is an arc of an ellipse with axes parallel to the coordinate axes, centered at $(0,d)$ and with semiaxis a,b in the plane $u = 0$. That is,

$$(7,29) \qquad \Lambda: \begin{cases} x = a \cos r \equiv \alpha(r) \\ y = d + b \sin r \equiv \beta(r) \\ u = 0 \equiv \kappa(r), \ a > 0, \ b > 0, \ d > 0, \ r \in I_r. \end{cases}$$

To complete Λ to a strip $E^o(r)$, we must find $p = \sigma(r)$, $q = \tau(r)$ such that

$$(7,30) \qquad \begin{cases} p\,\alpha' + q\beta' \equiv -pa \sin r + qb \cos r = 0 \\ F(x,y,u,p,q) \equiv p^2 + q^2 - \dfrac{a^2 \cos^2 r}{h^2} - \dfrac{(d+b \sin r)^2}{k^2} = 0. \end{cases}$$

From the first, we find

$$(7,31) \qquad q = p \cdot \frac{a}{b} \tan r \equiv \tau(r)$$

and by the second,

$$(7,32) \qquad p^2 = \frac{a^2k^2\cos^2 r + h^2(d + b \sin r)^2}{h^2k^2(1 + a^2tg^2r/b^2)} \equiv \sigma^2(r) \ .$$

Hence, there are two initial strips through Λ, which correspond to plus or minus the same absolute value of p; consequently also for q. Suppose I_r is open and contained in $(-\pi/2, 3\pi/2)$. Note that from continuity, it follows that, if p becomes zero, then it must change sign.

We may also assume I_r is arbitrary. But if $b = d$, then $r \neq -\pi/2 \pm 2m\pi$, since the origin $x = y = 0$ does not belong to our domain.

Assume $a = b = 0$, $d \neq 0$. Then Λ reduces to the point, $(0,d,0)$ and the first of (30) is identically satisfied. In this case, the "initial strip" becomes the Monge cone

$$\begin{cases} \dfrac{x-0}{p} = \dfrac{y-d}{q} = \dfrac{u-0}{\frac{d^2}{k^2}} \\ p^2 + q^2 = d^2/k^2 \end{cases}$$

that is,

$$(7,33) \qquad d^2x^2 + d^2(y-d)^2 - k^2u^2 = 0.$$

The initial strip $\overline{E}^{\circ}(r)$ is

$$(7,34) \qquad \overline{E}^{\circ}(r) = (0,d,0, \frac{d}{k}\cos r, \frac{d}{k}\sin r).$$

b) Consider the transversality condition. Its fulfillment has been already implied in the solving of (30) for p and q.

For the strip $E^{\circ}(r)$ given by (29-32), the transversality condition is

$$(7,35) \qquad \begin{vmatrix} \alpha' & \beta' \\ F_p & F_q \end{vmatrix} \equiv T(r) \equiv \begin{vmatrix} -a \sin r & b \cos r \\ 2p(r) & 2\,q(r) \end{vmatrix} \neq 0,$$

where $p(r)$ and $q(r)$ are given by (31) and (32).

It is obvious that the transversality condition is satisfied. Only for $r = 3\pi/2 \pm 2m\pi$ and $d = b$ might $T(r) = 0$, but then $x = 0 = y$, and this point is excluded.

Hence $E^{\circ}(r)$ is actually transverse and may be taken as an initial transverse strip for the posing of the Cauchy problem.

The case of the initial curve $\overline{E}^o(r)$ given by (34) is evidently a singular case. The integral surface obtained is the integral conoid at the point (0,d,0).

c) In the case of the equation we are considering in which the initial curve lies in the plane $u = u_o$, and only in this case, there is an important property having an interesting interpretation in the mentioned physical phenomena.

If and only if the supporting curve Λ lies on a plane $u = u_o$ is Λ orthogonal at each of its points, to the base characteristic through the same point and projection of the characteristic curve which originates at the corresponding element of E(r).

Actually, by virtue of the first equation of the characteristic system (3), one always has that along the characteristic strip the projection (p,q) of the normal (p,q,-1) coincides with the tangent (dx,dy) to the base characteristic curve. Therefore, for Λ to be orthogonal to (dx,dy), it is necessary and sufficient that

$$(7,36) \qquad \langle (\alpha'(r),\beta'(r)),(dx,dy)\rangle = \langle(\alpha',\beta'),(p,q)\rangle$$

$$= p\alpha' + q\beta' = 0.$$

But the strip condition requires

$$p\alpha' + q\beta' - \kappa' = 0.$$

Hence, the orthogonality condition is fulfilled if and only if $\kappa' = 0$, that is, $u = u_o$, q.e.d.

Note that $\alpha(r)$ and $\beta(r)$ are arbitrary, possibly even base characteristics. This is a property of every equation in which $q\,F_p = p\,F_q$.

The physical interpretation of this property is obvious, but observe it is true only when the initial curve Λ lies in a plane $u = u_o$.

d) Now, consider an initial curve Λ in the space $\{0;\ x,y,u\}$

$$(7,37a)\qquad \Lambda \begin{cases} \Lambda^* \begin{cases} x = \alpha(r) \\ y = \beta(r) \end{cases} \\ u = \kappa(r), \quad r \in I_r, \end{cases}$$

which may or may not support an initial strip

$$(7,37b)\qquad E^o(r) = (\ \Lambda,\ p = \sigma(r),\ q = \tau(r)\ ).$$

Consider also the matrix M,

$$(7,38)\qquad M = \begin{pmatrix} \alpha' & \beta' & \kappa' \\ F_p & F_q & pF_p + qF_q \end{pmatrix} \equiv \begin{pmatrix} \alpha' & \beta' & \kappa' \\ 2p & 2q & 2(p^2+q^2) \end{pmatrix}.$$

Since we are dealing with an integral strip, we must have

$$(7,39)\qquad \begin{aligned} & p\alpha' + q\beta' - \kappa' = 0 \\ & p^2 + q^2 = \frac{\alpha^2}{h^2} + \frac{\beta^2}{k^2} \end{aligned}$$

For $E^o(r)$ to be a singular initial strip, it is necessary and sufficient that

$$(7,40)\qquad q\alpha' - p\beta' = 0\ ,$$

together with the first equation of (39) yields

$$(7,41)\qquad p = \frac{\alpha'\kappa'}{\alpha'^2 + \beta'^2} \equiv \sigma(r),\quad q = \frac{\beta'\kappa'}{\alpha'^2 + \beta'^2} \equiv \tau(r).$$

Substituting in the second of (39), this becomes

$$(7,42)\qquad \kappa'^2 = (\alpha'^2 + \beta'^2)\left(\frac{\alpha^2}{h^2} + \frac{\beta^2}{k^2}\right).$$

Therefore, given a sufficiently smooth arbitrary curve in the plane $\{0;\ x,y\}$, say Λ^*, defined by the first two equations of (37), if one takes the function defined by (42) for $\kappa(r)$ in the same (37) and completes Λ with σ, τ given by (41), we obtain a singular initial strip $E^o(r)$, that is, (37a,b).

Now it turns out that this strip $E^o(r)$ is not only a singular strip, since it satisfys (40), but also a Monge strip. Actually, for this it is necessary and sufficient that the rank of M given in (38) be one, or, that

$$(7,43) \qquad 2\alpha'(p^2 + q^2) = 2\kappa' p \ ,$$

which in fact is satisfied due to (39) and (41), q.e.d.

At the conclusion of Article 2 we indicated how to construct a Monge strip. Let ds* be a differential element of arc on the curve Λ^*. We will show its relation to ds and dr where s is the auxiliary parameter of (3) and r the auxialiary parameter of (37).

According to the characteristic system and the given equation we have

$$(7,44) \qquad (ds^*)^2 \equiv dx^2 + dy^2 = (p^2 + q^2)ds^2 = \left(\frac{x^2}{h^2} + \frac{y^2}{k^2}\right) ds^2 \ .$$

This formula holds along any curve Λ^* satisfying the first two equations of (3) and whose strip $E^o(r)$ is integral. In particular, it holds for any Monge strip.

Let

$$(7,45) \qquad r = \rho(s^*)$$

be the relation of r with the arc length s*. We have

$$(7,46a) \qquad (ds^*) = dx^2 + dy^2 = (\alpha'^2 + \beta'^2)\ dr^2 \ ,$$

hence

$$(7,46b) \qquad \frac{ds^*}{dr} = \left(\alpha'^2 + \beta'^2\right)^{1/2} .$$

Now, according to (22), a necessary and sufficient condition that (37a,b) be a Monge strip is

$$(7,47) \qquad \frac{du}{ds^*} = \left(\frac{x^2}{h^2} + \frac{y^2}{k^2}\right)^{1/2} ,$$

hence, by (46)

$$(7,48) = (7,42) \quad \kappa'(r)\, dr = du = \left(\frac{\alpha^2}{h^2} + \frac{\beta^2}{k^2}\right)^{1/2} \left(\alpha'^2 + \beta'^2\right)^{1/2} dr.$$

Thus, we have again shown that $E^O(r)$ given by (37a,b) is a Monge strip if and only if $u = \kappa(r)$ is given by (42) and σ, τ by (41).

We finish by remarking that $E^O(r)$, which is a Monge strip, will be a characteristic strip if and only if Λ^* is a base characteristic.

5. a) Integral Surfaces. Suppose we are given a transverse initial strip $E^O(r)$ such as that in (29-32) and we substitute these five functions in (9) for $E^O=(x_o,y_o,u_o,p_o,q_o)$ respectively; then one has the set of base characteristics, projections of the characteristics in the space $\{O;\ x,y,u\}$, which generate the integral surface. The base characteristics represent the paths along which the vibrations of the membrane or light rays propagate. The characteristic curves in the space $\{O;\ x,y,u\}$ represent the trajectories since they relate the path with the time u.

Suppose $h = k$. Then substituting $E^O(r)$ for E^O in (9) and (18) we obtain the equations of the integral surface as functions of the parameters s and r. To obtain the surface in explicit form $u = \psi(x,y)$, we must eliminate s and r between the three mentioned equations.

b) The integral conoid at the point $(0,y_o,0)$ is particularly interesting.

In this case, the initial degenerated strip is $\overline{E^o(r)}$, given by (34) setting $d = y_o$.

The base characteristics are given by (11), where r and y_o are the same as those of the initial strip (34). This follows from (10) setting $d = y_o$.

The integral conoid at $(0,y_o,0)$ is given by (11) and (18), or better (19). Since (10) is satisfied, they yield

$$(7,49) \qquad u = \frac{xy}{k \cos r} \quad .$$

Now, we eliminate s between (11), and solve the resulting equation for r. Substitute r in (49). The elimination is done, resulting in (11b). Now, eliminating cos r between (11b) and (49), we obtain

$$(7,50) \qquad 4k^2u^2 = (x^2 + y^2 + y_o^2) - 4y_o^2y^2 ,$$

that is,

$$(7,51) \qquad 4k^2u^2 = (x^2 + (y - y_o)^2)(x^2 + (y + y_o)^2) .$$

This surface has two conic points, at $x = 0$, $y = \pm y_o$. If we consider only the portion of the surface generated by the characteristics passing through $(0,y_o,0)$, that is, only for $y > 0$, then the surface consists of two sheets separated by the vertex $(0,y_o,0)$ and symmetric with respect to the plane $u = 0$. The sheet with $u > 0$ is the one looking at the future, and that with $u < 0$ looks at the past.

c) Setting $u = u_1 \equiv$ constant, that is, for time $u = u_1$, (51) gives the wave front assuming that at time

0, an impulse was given to the membrane at $(0, y_o, 0)$; it gives the wave front of the light rays of an instantaneous sh originating at $(0, y_o, 0)$ at time $u = 0$. See figure 2.

Now, for $u = u_1 \equiv$ constant, the curves (51) are the sini ovals, that is, the locus of points such that the duct of the distances to two fixed points $(0, y_o, 0)$, $-y_o, 0)$ is constant and equals $2\, u_1 k$.

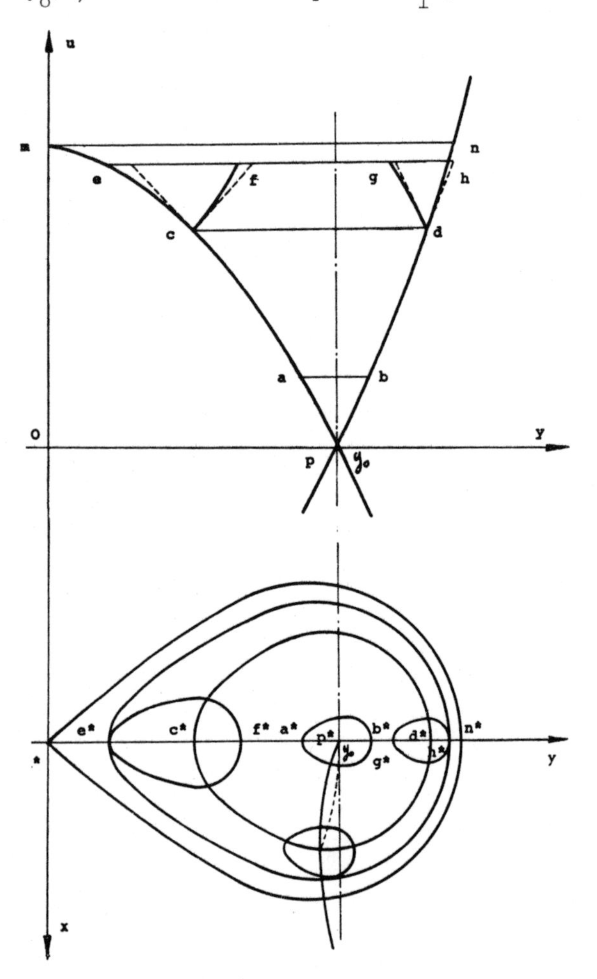

2. Integral Conoid with Vertex at P.
Profile, Wave Fronts and Huyghens' Construction.

The Cassini ovals or wave fronts are orthogonal to the characteristic hyperbolas passing through $(0,y_o,0)$. This result was to be expected since it is a consequence of the mentioned (article 4,b) orthogonality of the initial curve with the characteristic when the initial curve lies in a plane $u \equiv$ constant. Moreover, one can easily verify this taking derivatives of (51) with respect to x. Assuming u is constant, and changing $\frac{dy}{dx}$ to $-\frac{dx}{dy}$, we obtain

$$(7,52) \qquad (x^2 + y^2 + y_o^2)y' - (x^2 + y^2 - y_o^2)y = 0,$$

which is the same differential equation obtained by eliminating tg r between (12) and its derivative with respect to x.

It is interesting to note that in the Euclidean plane $\{0; x,y\}$, the characteristic curve does not reach any point (x,y) such that $y \leqq 0$.

d) Considering the surface obtained by cutting the integral conoid at $(0,y_o,0)$ by planes $u \equiv$ constant or other transverse surfaces, it is easy to get an idea of the integral surface through any transverse initial strip.

Also, if the initial strip is a characteristic strip, one sees there are infinitely many integral surfaces through it.

Suppose we are given a generic transverse initial strip and an integral surface through it. Then the base characteristics are a 1-parameter system of hyperbolas contained in (12) and in general, have an envelope, call it Λ^*. The curve Λ lying in the integral surface which projects on Λ^* will be a Monge curve and singular curve. A kind of edge of regression; at its points the integral surface cannot have continuous second derivatives. In the case of the crystal, such a curve is called a caustic.

Note that through any transverse initial curve Λ, there always originates 4 sheets constituting two integral surfaces crossing at Λ. Of the four sheets, two look to the future and two look to the past.

6.a) Consider the general application to geometric optics of equation (1) and its characteristic strips and integral surfaces. There is a suggested connection with Fermats' principle of geometric optics according to which light travels from focus to object by the shortest (longest) path in time.

Let us formulate the conjecture which will lead us to consider Fermat's variation principle. Once formulated, its proof will be easy. We could take the y-axis through the focus and the moment of the light flash as the origin of time. Then $(0, y_o, 0)$ would be the focus in the space-time crystal and there would be no loss of generality. For the moment, assume in general that $P_1 = (x_1, y_1, u_1)$ is the focus and $P_2 = (x_2, y_2, u_2)$, $u_2 \geqq u_1$, is the object at which the ray arrives at the undetermined time u_2.

Let $d\bar{s}$ be the differential element of arc of trajectory,

$$(7,53) \qquad d\bar{s}^2 = (dx^2 + dy^2) + du^2 = (ds^*)^2 + du^2 ,$$

where ds^* is the differential element of arc of the path already introduced in (22) and (46). The differentials ds^* and du are necessarily related by the refraction index of the crystal due to (22).

$$(7{,}54) = (7{,}22) \qquad \frac{ds^*}{du} = \left(\frac{x^2}{h^2} + \frac{y^2}{k^2} \right)^{-1/2}.$$

That implies that the possible trajectories of light rays going from P_1 to P_2 are the Monge curves which originating at (x_1, y_1) at the instant u_1, arrive at (x_2, y_2) at the undetermined instant $u_2 \geqq u_1$. Fermat's principle implies that among all trajectories, the real light ray will take that one for which $u_2 - u_1$ is extremal, that is, the path taking the least or most time.

This is equivalent to saying among all Monge curves or <u>focal curves</u> given by (37a), (42) or (48), the light rays pass over those which make extremal the integral

$$(7{,}55) \qquad J_r\,[\alpha, \beta, \kappa] \equiv \int_{(x_1, y_1)}^{(x_2, y_2)} du.$$

Applying (54) and (46), or directly, (42) or (48), we obtain

$$(7{,}56) \qquad J_r\,[\alpha, \beta] \equiv \int_{(x_1, y_1)}^{(x_2, y_2)} \left(\frac{\alpha^2}{h^2} + \frac{\beta^2}{k^2} \right)^{1/2} \left({}'^2 + \beta'^2 \right)^{1/2} dr.$$

Here we assume that equation (1) is a sufficiently exact model of geometric optics. Then setting $h = k$, it can be expected that the extremals of J are precisely the equilateral hyperbolas given by (14), where (x_o, y_o) is replaced by (x_1, y_1), and one calculates C_o, if possible, under the condition the hyperbola goes through (x_2, y_2).

Now, considering only the plane $\{0;\, x, y\}$ we can obtain the same functional J_r if we introduce the speed $v(r)$ with which the ray runs over the possible path. By (22) and (46),

$$du = \frac{ds^*}{v} = \frac{ds^*}{\tan \omega} = \left(\frac{\alpha^2}{h^2} + \frac{\beta^2}{k^2} \right)^{1/2} \left(\alpha'^2 + \beta'^2 \right)^{1/2} dr,$$

which yields the same functional J_r. This is what we set out to show.

To find the extremals of $J_r[\alpha,\beta]$, set $x = \alpha(r) \equiv r$ and $y = \beta(r) \equiv \beta(x)$, then (56) becomes

$$(7,57) \qquad J_x[\, y \,] = \int_{x_1}^{x_2} \left(\frac{x^2}{h^2} + \frac{y^2}{k^2} \right)^{1/2} \sqrt{1 + y'^2} \; dx.$$

Setting

$$\left(\frac{x^2}{h^2} + \frac{y^2}{k^2} \right)^{1/2} \sqrt{1 - y'^2} \equiv g(x,y,y'),$$

the extremals are the integral curves of Euler's equation

$$\frac{d}{dx} \frac{\partial g}{\partial y'} - \frac{\partial g}{\partial y} = 0 .$$

Carrying out the calculations, we obtain the second order differential equation

$$(7,58) \qquad \left(\frac{x^2}{h^2} + \frac{y^2}{k^2} \right) \frac{d^2y}{dx^2} + \frac{x}{h^2} \left(\frac{dy}{dx} \right)^3 - \frac{y}{k^2} \left(\frac{dy}{dx} \right)^2 + $$

$$+ \frac{x}{h^2} \frac{dy}{dx} - \frac{y}{k^2} = 0 .$$

Setting $h = k$, this equation is the same as (16), and therefore the characteristic curves are precisely those which make the functional J extremal, q.e.d.

b) Consider in particular the integral conoid (52) generated by the light rays originating with a flash at the focus $(x_1, y_1, u_1) = (0, y_o, 0)$. Consider the two Cassini ovals corresponding to $u = u_2$, $u = u_3$, $u_3 > u_2$. The base characteristics are orthogonal to these ovals. Measuring in time units the distance between the ovals along the characteristics, we see that the ovals are concentric circles which correspond to the wave fronts at the instants u_2, u_3. In other words, if on the crystal instead of using a Euclidean compass we use a compass obeying the proper (exercise 8) Riemannian metric dt, which is equal to du,

$$(7,59) \qquad (dt)^2 = \left(\frac{x^2}{h^2} + \frac{d^2}{k^2} \right) (ds^*)^2, \quad (ds^*)^2 = dx^2 + dy^2 ,$$

then placing one tip at $(0, y_o, 0)$ the other traces out a Cassini oval. In the plane of the crystal provided with the metric dt, the base characteristics are geodesics of the plane, that is, the radii of the mentioned circles.

This is a general property of all wave fronts corresponding to the same integral surface, that is, belonging to the same physical phenomena. Only instead of speaking of concentric circles, we must speak of parallel curves. In the Euclidean plane two curves are said to be parallel when the distance from any point of one to the other measured along a common perpendicular (which always exists) is constant. For instance, the system of <u>evolvents</u> of the same evolute, which is the common envelope of the common normals to the <u>evolvents</u>

of the system, is a system of parallel curves.

Suppose an integral surface Σ has been constructed by successive flashes along a curve which supports a transverse initial strip. Let C_1, C_2 be two transverse curves obtained by cutting Σ by the planes $u = u_1, u = u_2$ respectively. Consider the paths C_1^*, C_2^* in the crystal which have been wave fronts at times u_1 and u_2. The base characteristics along which the light rays propagate form a system orthogonal to C_1^*, C_2^*, and the distance between both curves along a characteristic, measured in the metric dt, is constant. Therefore C_1^*, C_2^* are parallel curves in the metric dt. If with centers on C_1^* we draw circles of constant radius u_2-u_1 in the metric dt, and then draw the envelope of the circles thus obtained, we get the curve C_2^*. The mentioned circles are projections of the intersections of the plane $u = u_2$ with the integral conoids at the points of C_1. This process is Huyghens' construction of successive wave fronts.

If the base characteristics orthogonal to C_1^* and C_2^* have an envelope, it is a caustic which may be any curve of the plane $\{0; x,y\}$, although considered in $\{0; x,y,u\}$ it must be a focal curve.

We introduce the Riemanian metric

$$d\ell^2 = -\left(\frac{x^2}{h^2} + \frac{y^2}{k^2}\right)(dx^2 + dy^2) + du^2$$

in the space $\{0; x,y,u\}$. This metric is naturally associated with the differential equation (2) and enjoys invariance properties with respect to systems called inertial. With this metric, the trajectories evidently are curves of null length. For a habitant of the crystal world, the material universe, or space time for which $d\ell^2 > 0$, reduces to the interior of the integral conoid with vertex at (x,y,u) where (x,y) is the point he occupies and u is the time on his watch. The metric measures the length between two infinitely close events. The integral conoid contains all events simultaneous with the vertex.

That these considerations are easily extended to a space $\{O; x,y,z,u\}$ in the domain $\{(x,y,z)|x^2 + y^2 > 0\}$ follows from the method of descent.

Although for the physical interpretation we have appealed to the physical phenomena of light propagation in a crystal, the results may as well apply, but without the same precision, to the vibrating membrane.

Exercises

1. Suppose we are given the equation of light rays

$$p^2 + q^2 = 1 .$$

a) Find a characteristic curve through the points (0,0,0) and (3,4,u*) where u* is real.

b) Find a characteristic through (0,0,3) and $(3,4,\overline{u})$ where $\overline{u}$ is real.

c) Explain the relation between the results of a) and b).

2. Let r be a straight line of the plane $\{O; x,y\}$, M a point of r and C* the base characteristic (hyperbola) of equation (1) with h = k that is tangent to r at M.

Let γ be the circle centered at M and of radius $\overline{MO}$, and let P, Q be the points of intersection of γ with r. Show that $\overline{OP}$ and $\overline{OQ}$ are the axes of C*.

3. Deduce equation (12) starting from (26). [Consider

$$\frac{ha}{2}\frac{\partial}{\partial a}\log\left[x + (x^2 - ah^2)^{1/2}\right] = \frac{h}{4} - \frac{hx}{4}(x^2 - ah^2)^{-1/2},$$

and changing the constants obtain (8). Setting h = k, rationalizing and changing constants again, obtain (12).]

4. Suppose h = k in (1), and $\overline{OP} = y_o$, determine the equations of the curves m e c a p and p b d h n of figure 2.

5. Given the equation $p^2 + q^2 = 1$, let μ be a Monge curve such that

$$p = \cos(ax + b)$$

where a,b are real.

a) Find the projections of μ on the planes {O; x,y} and {O; x,u}.

b) Verify that μ forms a constant angle $\pi/4$ with the u-axis.

c) Find conditions on a and b for μ to be a characteristic curve.

6. Given the equation $p^2 + q^2 = f(x,y)$ and the paraboloid

$$2u = (x - 3)^2 + (y - 3)^2,$$

which is an integral surface of the given equation, consider the element $E^o = (x_o, y_o, u_o, p_o, q_o)$ belonging to the paraboloid such that $x_o = y_o = 1$.

a) Define the strip B which passes through E^o and is a solution of the characteristic system.

b) Determine whether or not B is integral.

c) Find the maximal continuation of B.

d) Explain the position of B relative to the paraboloid.

7. Given the equation $p^2 + q^2 = 1$, and the curve

$$L: \begin{cases} y = 1 \\ u^2 = x^2 + 1 \end{cases};$$

a) Find the equation of the integral surface S through L.

b) Knowing that $P = (3/4, 1, 5/4) \in L$ and $M = (a, b, 10)$ belong to the same characteristic C lying in S, determine the values of a,b.

c) Find the value of $\hat{u}$ knowing $\hat{M} = (a, b, \hat{u})$ lies in the characteristic curve supporting the characteristic strip passing through the element $E^o = (3/4, 1, 6, 3/5, q)$.

d) Give an interpretation of the results in terms of geometric optics in a homogeneous plane crystal.

8. Verify that the metric

$$dt^2 = \left(\frac{x^2}{h^2} + \frac{y^2}{k^2}\right)(dx^2 + dy^2) \equiv g_{ij}\, dx^i\, dx^j$$

defined by (59) is properly Riemannian. [Setting h = k, we obtain $R_{1212} = -3/4\ k^2$.]

Chapter III

EQUATIONS WITH MORE THAN TWO INDEPENDENT VARIABLES

§ 8. The Cauchy Method

1. Characteristic System and Strips. - We consider the equation

$$F(x,u,p) = 0, \; x = (x_1, \ldots, x_n), \; p = (p_1, \ldots, p_n),$$

$$(8,1) \qquad p_i \equiv \frac{\partial u}{\partial x_i},$$

$$(x,u,p) \in D^{2n+1} \subset E^{2n+1}, \; x \in D^n \subset E^n, \; (x,u) \in D^{n+1}$$

$$F \in C^2(D^{2n+1}), \; |\hat{F}_1| + \ldots + |\hat{F}_n| \neq 0, \; \hat{F}_i \equiv \frac{\partial u}{\partial p_i}.$$

We assume F is real and the 2n-dimensional manifold of solutions of $F = 0$ in a neighborhood of any point of D^{2n+1} is also real.

We say $u = \phi(x)$ is a solution, or integral, or integral hypersurface of (1) for $x \in G$, a domain in $\{0; x\}$, if

$$(8,2) \quad \begin{array}{l} 1) \; \phi \in C^2(G) \\ 2) \; F(x,\phi(x), \frac{\partial \phi(x)}{\partial x}) = 0, \; x \in G, \\ \frac{\partial \phi(x)}{\partial x} \equiv (\phi_1, \ldots, \phi_n), \; \phi_i = \frac{\partial \phi}{\partial x_i} \end{array}$$

We assume ϕ is real.

We shall use the following notation. For any scalar function v depending on x, u or p_i, assuming all arguments vary independently, the partial derivative will be expressed by

$$(8,3a) \qquad \frac{\partial v}{\partial x_i} \equiv v_i, \; \frac{\partial v}{\partial u} \equiv v_u, \; \frac{\partial v}{\partial p_i} \equiv \hat{v}_i.$$

$$\frac{\partial v}{\partial x} \equiv \frac{\partial v}{\partial (x)} \equiv (v_1, \ldots, v_n), \quad \frac{\partial v}{\partial p} \equiv (\hat{v}_1, \ldots, \hat{v}_n).$$

If v is a vector, $v = (v_1, \ldots, v_m)$, then we use

$$\frac{\partial v}{\partial x_i} \equiv v_{,i} \equiv (v_{1,i}, \ldots, v_{m,i})$$

$$(8,3b) \qquad \frac{\partial v}{\partial x} \equiv \frac{\partial (v)}{\partial (x)} \equiv \begin{pmatrix} v_{1,1} & v_{1,2} & \cdots & v_{1,n} \\ \cdots & \cdots & \cdots & \cdots \\ v_{m,1} & v_{m,2} & \cdots & v_{m,n} \end{pmatrix}.$$

The derivatives represented by subindices always indicate partial derivatives with respect to the variable specified by the index, assuming all other arguments are independent of the specified variable. Otherwise, we speak of composed partial derivatives and shall use the notation of Jacobi. Thus

$$\frac{\partial \phi(x,u)}{\partial x_i} \equiv \phi_i + \phi_u \, p_i \neq \phi_i \, .$$

A set of (2n + 1) values $E^o = (x^o, u_o, p^o) \in D^{2n+1}$ is called an element. It is said to be integral if $F(E^o) = 0$. Also, we call (x^o, u_o) the supporting point of E^o. It is clear that an element can be interpreted as a point in D^{n+1} and a plane through this point whose normal has direction parameters $(p_1^o, \ldots, p_n^o, -1)$. An m-dimensional manifold B, $1 \leq m \leq m - 1$, such that to each point $M \in B$ there is associated a hyperplane P(M) in such a way that this manifold is converted to an m-parameter system of elements constituting a strip, is called a hyperstrip with m-dimensional support. A necessary and sufficient condition that this system constitutes a strip is it satisfies the strip condition, that is, for each point M, the hyperplane P(M) contains the m-dimensional tangent plane to B at M. This is represented analytically as follows:

Let Γ be an m-dimensional manifold immersed in D^{n+1},

$$(8,4a) \qquad \Gamma : \begin{cases} x = \xi(r) \\ u = \zeta(r) , \ r = (r_1, \ldots, r_m) \ \varepsilon \ D^m . \end{cases}$$

We complete Γ to an m-parameter system of elements by adjoining equations

$$(8,4b) \qquad p = \pi(r), \ \pi \ \varepsilon \ C^1(D^m).$$

This system will be a strip with m-dimensional support if

$$(8,4c) \qquad \sum_{k=1}^{n} \pi_k \ \xi_{k,i} = \zeta_i \ , \ r \ \varepsilon \ D^m \ , \ i = 1,2, \ldots , m,$$

where the subindex i indicates partial derivation with respect to r_i. Moreover, we consider exclusively manifolds Γ which are explicit in the variable u. Hence, we insist that the closure of the projection Γ^* of Γ on $\{0;x\}$ is simple.

A strip is said to be <u>integral</u> if all its elements are integral. It is said to be a <u>hyperstrip</u> if it has (n - 1) - dimensional support.

Let us now consider the following ordinary differential system of 2n equations, not counting s,

$$(8,5) \qquad \frac{dx_1}{\hat{F}_1} = \ldots = \frac{dx_n}{\hat{F}_n} = \frac{du}{p_1\hat{F}_1 + \ldots + p_n\hat{F}_n} =$$

$$= \frac{dp_1}{-F_1 - p_1F_u} = \ldots = \frac{dp_n}{-F_n - p_nF_u} = ds.$$

This system is called the characteristic system of $F = 0$. Given any initial element $E^o = (x^o, u_o, p^o)$, there exists a unique solution of the characteristic system through it. This system can be considered as a 1-parameter system of elements $E(s;\ s_o,\ E^o)$ such that $E(s_o;\ s_o,\ E^o) = E^o$. From the first n-equations, it follows that this 1-parameter system of elements is a strip supported by a curve (1 - dimensional manifold), since along this curve, the strip condition

$$p_1(s) \cdot x_1'(s) + \dots + p_n(s) \cdot x_n'(s) = u'(s),\ s \in I_s\ ,$$

is fulfilled. In this equation, the primes mean derivatives with respect to s. Hence, in the following, we may speak of strip solutions of the characteristic system. It is easily verified that along any solution of the characteristic system, that is, as s varies,

$$\frac{dF}{ds} = 0\ ,\ s \in I_s\ ,$$

and therefore $F(x,u,p)$ is a first integral of the system. If the strip solution is determined by an initial element E^o, then it will be an integral strip if and only if $F(E^o) = 0$.

The integral strip solutions of the characteristic system are called characteristic strips, and their supporting curves characteristic curves. Since there are 2n equations, the solution set depends on 2n arbitrary parameters, and since $F(E^o) = 0$ imposes a relation among the same, it follows that the system of characteristic strips depends on (2n -1) arbitrary parameters.

Finally, we give a definition of a <u>transverse</u> <u>initial</u> <u>hyperstrip</u> $E^o(r)$. It is an integral strip with (n-1)-dimensional support Λ,

$$(8,6a)\quad E^o(r): \begin{cases} \Lambda: \begin{cases} x = \alpha(r),\ r = (r_1,\dots,r_{n-1}) \in J^{n-1} \\ u = \kappa(r);\ \alpha,\kappa \in C^2(J^{n-1}) \end{cases} \\ p = \sigma(r),\ \sigma \in C^1\ (J^{n-1}). \end{cases}$$

In the general definition of a strip, it is assumed that J^{n-1} is a domain of $\{0;\ r\}$ and that for all $r \in J^{n-1}$, the rank of $\frac{\partial\alpha}{\partial r}$ is (n-1), and moreover, the closure of the projection Λ^* of Λ on $\{0;x\}$ is simple.

The functions α, κ, σ must satisfy

$$(8,6b)\quad \begin{cases} F(\alpha(r),\ \kappa(r)) = 0 \\ \sigma_1\ \alpha_{1,i} + \dots + \sigma_n\ \alpha_{n,i} - \kappa_i = 0\ , \\ r \in J^{n-1},\ i = 1,\ \dots\ ,\ n-1\ . \end{cases}$$

The first expresses that the hyperstrip $E^o(r)$ is integral, while the other (n-1) are the strip conditions.

This initial hyperstrip must be transverse, that is, to the characteristic curves. Just as in the 2-variable case, we impose the transversality condition in the hyperplane $\{0;x\}$ between Λ^* and the <u>base</u> <u>characteristics</u> which are projections on $\{0;x\}$ of the characteristic curves. The transversality condition is expressed by

$$(8,6c)\qquad \Delta = \begin{vmatrix} \alpha_{1,1} & \cdots\cdots\cdots & \alpha_{n,1} \\ \cdot & \cdots\cdots\cdots & \cdot \\ \alpha_{1,n-1} & \cdots\cdots\cdots & \alpha_{n,n-1} \\ F_1(\alpha(r),\kappa(r),\sigma(r)) & \cdots & F_n \end{vmatrix} \neq 0,\ r \in J^{n-1}.$$

Note, that although this condition is predicated of Λ^*, it also depends on the functions κ, σ. The determinant Δ is precisely the Jacobian of (6b) with respect to σ_i. Therefore, given a fixed element $E^* \equiv (\alpha(r^*), \kappa(r^*), \sigma(r^*)) = E^o(r^*)$ satisfying (6b) and (6c), the functions σ are uniquely determined by (6b)in a neighborhood of r^*, assuming α and κ are known, that is, assuming Λ is known; in the same way as in the two variable case. We can now state and solve the Cauchy problem.

2. <u>The Cauchy Problem</u>.- Given the transverse initial hyperstrip (6), find, in a domain G of $\{0;x\}$, an integral surface $u = \phi(x)$, $x \in G$, of $F = 0$ such that $\Lambda^* \subset G$ and

$$(8,7)\qquad \phi(\alpha(r)) = \kappa(r),\ \phi_i(\alpha(r)) = \sigma_i(r),\ i = 1,\ldots,n.$$

As in the 2 variable case, we solve the problem with two theorems.

<u>Theorem 8.1</u> <u>Let</u> E^* <u>be an element of the given integral surface</u> Σ: $u = \phi(x)$, $x \in G$. <u>Let</u> $E(s;s_o,E^*)$ <u>be the characteristic strip such that</u> $E(s_o;s_o,E^*) = E^*$. <u>Then the strip</u> $E(s;s_o,E^*)$ <u>lies in</u> Σ.

The proof is left as Exercise 1, since one needs only to generalize the proof given of Theorem 5.1. In fact, due to vectorial notation, the proof is simpler.

As in the 2-dimensional case, it is clear that this theorem implies the uniqueness of the solution of the Cauchy problem in the following specified sense.

Theorem 8.2 The Cauchy problem is such that:

I) There is at most one solution.

II) There exists a solution $u = \phi(x)$, $x \in G$.

III) The dependence of ϕ on the Cauchy data κ, σ is continuous.

Here, as in Theorem 5.2, we omit the proof and precise formulation of (3), but it may be given in a fashion analogous to Theorem 3.2.

Proof I) The meaning of uniqueness is the following. If ϕ_1, ϕ_2 are solutions in G_1, G_2 respectively, then

$$\phi_1(x) = \phi_2(x) \text{ , for all } x \in G = G_1 \cap G_2 \text{ ,}$$

where G must necessarily be a domain containing Λ^*. The proof is an obvious consequence of Theorem 5.1.

II) There exists a solution $u = \phi(x)$, $x \in G$, which is given parametrically by

$$
(8,8) \quad \begin{cases} x = \xi(s;s_o,E^o(r)), & \xi(s_o;s_o,E^o(r)) = \alpha(r) \\ u = \zeta(s;s_o,E^o(r)), & \zeta(s_o;s_o,E^o(r)) = \kappa(r), \end{cases}
$$

$$
r \in J^{n-1}, \; \delta_1(r) < s-s_o < \delta_2(r), \; \delta_1 > 0, \; \delta_2 > 0,
$$

where $x = \xi(s;s_o,E^*)$, $u = \zeta(s;s_o;E^*)$, $p = \pi(s;s_o,E^*)$ are the integral of the characteristic system (5), which for $s = s_o$, pass through E^*. Hence, the surface passes through Λ, or better, contains the initial hyperstrip $E^o(r)$, as required by (7).

The surface given in (8) is of class two. Actually, due to (1), the characteristic system may be put in normal form, and it is regular since all functions of the second members are of class one. Hence, according to a well known regularity theorem, ξ, ζ, and π are also of class one. This, and the transversality condition (6,c), imply that, as in the case $n = 2$, $\phi \in C^2(G)$.

We omit the proof that the surface (8) satisfies

$$
F(\xi(s;r),\zeta(s;r),\pi(s;r)) = 0, \; r \in J^{n-1}, \; \delta_1 < s-s_o < \delta_2,
$$

which can be given by generalizing the proof of Theorem 5.2.

Certainly, the surface (8) can be put in explicit form. From the transversality condition (6c) follows

$$
(8,9) \qquad \frac{\partial(\xi)}{\partial(s,r)} \neq 0 \; , \; s = s_o \; , \; r \in J^{n-1} \; ,
$$

and therefore one can solve $x = \xi(s;s_o,E^o(r))$ for $s = s_o$ and $r \in J^{n-1}$. By the usual continuity argument, one can then also solve the system for s and r as functions of x in a neighborhood of Λ^*. Substituting $s(x)$, $r(x)$ in the last of (8), we obtain the surface $u = \phi(x)$ in explicit form, defined in a neighborhood G of Λ^*, q.e.d.

3. _The singular case, $\Delta = 0$._ We examine the case $\Delta = 0$ in (6c).

We define a _characteristic strip with m-dimensional support_, $1 \leqq m \leqq n - 1$, as an integral strip with m-dimensional support which can be considered as being generated by characteristic strips through a transverse initial integral strip with (m-1)-dimensional support.

In speaking of a characteristic strip without specifying the dimension of the support, it is understood that $m = 1$, that is, we are dealing with a particular integral of the characteristic system. One could say that an integral surface is a characteristic strip with n-dimensional support with the understanding that the associated hyperplane at each point is the tangent hyperplane. This is precisely the content of Theorem 8.1.

Another definition, obviously equivalent, is the following: an integral strip B with m-dimensional support is a characteristic strip with m-dimensional support, if for each element $E^o \in B$, the characteristic strip through E^o lies in B. Consider the analytic formulation of the first definition.

Let

$$(8,10) \qquad B : \mathbf{x} = \alpha(r),\ u = \kappa(r),\ p = \sigma(r),$$
$$r = (r_1, \ldots, r_m) \in I^m .$$

be a strip with m-dimensional support. The integral strip B is characteristic with m-dimensional support, if there exists a change of variables of class two

$$r \to (s,t),\ t = (t_1,\dots,t_{m-1}),\ r = \rho(s,t),\ \rho \in C^2(J^m),$$

$$\left| \frac{\partial(\rho)}{\partial(s,t)} \right| \neq 0 ,$$

such that for the new equations of B one has

$$(8,11a) \qquad x = \alpha(\rho(s,t)) = \xi(s,t),\ u = \zeta(s,t),\ p = \pi(s,t) ,$$

and

$$(8,11b) \qquad \frac{dx_i}{ds} = \hat{F}_i,\ \frac{du}{ds} = \sum_{h=1}^{n} p_h \hat{F}_h,\ \frac{dp_i}{ds} = -F_i - p_i F_u,$$

$$i = 1,\dots,n,\ (s,t) \in J^m,$$

where it is understood that these equations are identically satisfied in s and in $t = (t_1,\dots,t_{m-1})$ considered as a parameter.

We leave as Exercise 2 the proof that the strip conditions are invariant.

Therefore, to construct a characteristic strip with m-dimensional support $B(r)$, it suffices to have an integral strip with (m-1)-dimensional support $B_o(\bar{r})$ transverse to the characteristic strips through its elements, and to take it as an initial strip. Let

$$(8,12) \qquad x = \bar{\alpha}(\bar{r}),\ u = \bar{\kappa}(\bar{r}),\ p = \bar{\sigma}(\bar{r}),$$

$$\bar{r} = (\bar{r}_1,\dots,\bar{r}_{m-1}) \in \bar{J}^{m-1},$$

be the equations of an integral strip with (m-1)-dimensional support $B^{o}(\bar{r})$. The transversity condition is that the matrix N,

$$(8,13)\quad N = \begin{pmatrix} \bar{\alpha}_{1,1} & \cdots\cdots\cdots & \bar{\alpha}_{n,1} \\ \cdots\cdots\cdots & \cdots\cdots\cdots & \cdots\cdots\cdots \\ \bar{\alpha}_{1,m-1} & \cdots\cdots\cdots & \bar{\alpha}_{n,m-1} \\ \hat{F}_1(\bar{\alpha}(\bar{r}), \bar{\kappa}(\bar{r}), \bar{\sigma}(\bar{r})) & \cdots & \hat{F}_n \end{pmatrix}$$

is of rank m for all $\bar{r} \in \bar{J}^{m-1}$.

The equations of the characteristic strip B(r) will be

$$(8,14)\quad B(r) \begin{cases} x = \xi(s;s_o,B^{o}(\bar{r})), & \bar{r} \in \bar{J}^{m-1}, \delta_1(\bar{r})<s-s_o<\delta_2(\bar{r}) \\ u = \zeta(s;s_o,B^{o}(\bar{r})), & r \equiv (s,\bar{r}) \in J^{m}, \\ p = \pi(s;s_o,B^{o}(\bar{r})), & \end{cases}$$

where ξ,ζ,π are the general integral of (5) passing through $B^{o}(\bar{r})$ for $s = s_o$. This strip with (m-1)-dimensional support is defined in a domain J^m which is a neighborhood of $\bar{J}^{m-1}$ in $\{0;s,\bar{r}\}$.

From the given definition, we deduce a necessary and sufficient condition for a strip B(r) with m-dimensional support to be characteristic. Let B(r) be given by (10).

Assertion 8.1. _For_ B(r) _to be a characteristic strip with_ m-_dimensional support, it is necessary and sufficient that for all_ $r \in I^m$, _the matrix_ Q

$$
(8,15)\quad Q = \begin{pmatrix}
\alpha_{1,1} & \cdots & \alpha_{n,1} & \kappa_1 & \sigma_{1,1} & \cdots & \sigma_{n,1} \\
\cdots & \cdots & \cdots & \cdots & \cdots & \cdots & \cdots \\
\alpha_{1,m} & \cdots & \alpha_{n,m} & \kappa_m & \sigma_{1,m} & \cdots & \sigma_{n,m} \\
\hat{F}_1(\alpha(r),\kappa(r),\sigma(r)) & \cdots & \hat{F}_n & \sum_{h=1}^{n} \sigma_h \hat{F}_h & -F_1-\sigma_1 F_u & \cdots & -F_n-\sigma_n F_u
\end{pmatrix}
$$

$$
\kappa_i = \frac{\partial u}{\partial r_i} \quad ,
$$

be of rank m, and that

$$
(8,16)\qquad F(\alpha(r),\kappa(r),\sigma(r)) = 0.
$$

Proof. From the definition, there is a change of variables $r = \rho(s,t)$ such that

$$
\frac{dx_i}{ds} = \sum_{k=1}^{n} \alpha_{i,k} \cdot \rho_{k,s} = \hat{F}_i \ , \ i = 1,..,n
$$

$$
\frac{du}{ds} = \sum_{k=1}^{n} \kappa_k \cdot \rho_{k,s} = \sum_{h=1}^{n} \sigma_h \hat{F}_h
$$

$$
\frac{dp_i}{ds} = \sum_{k=1}^{n} \sigma_{i,k} \cdot \rho_{k,s} = -F_i - \sigma_1 F_u \ .
$$

This says precisely that the last row of Q is a linear combination of the first m. Hence $r(Q) \leqq m$. But, from the definition of strip, at least one minor of the first m rows and first n columns of Q is non-singular. Hence $r(Q) \geqq m$. Therefore $r(Q) = m$. Condition (16) expresses the fact that the strip is integral. These conditions are evidently sufficient also, although perhaps not all independent, and the assertion is proven. An important property of characteristic strips with m-dimensional support is the following: if two characteristic strips with dimensions of supports m_1, m_2 respectively, intersect, the intersection is also a characteristic strip. The following assertion is obvious.

Assertion 8.2. Let B_1, B_2 be characteristic strips and E^o an element such that $E^o \in B_1 \cap B_2$.

Then if $E(s;s_o,E^o)$ is the characteristic strip through E^o, also

$$E(s;s_o,E^o) \subset B_1 \cap B_2 \quad .$$

By a characteristic hyperstrip, we mean a characteristic strip with (n-1)-dimensional support. By a singular initial hyperstrip we mean an integral one with (n-1)-dimensional support such as $E^o(r)$ given by (6a,b), but such that instead of (6c), we assume

$$(8,17) \qquad \Delta = 0 \ , \ r \in J^{n-1} \quad .$$

Assertion 8.3. Let Σ: $u = \phi(x)$, $x \in G$, be an integral surface and $E^o(r)$,

$$(8,18a)=(8,6a) \quad \begin{cases} E^o(r) : \begin{cases} \Lambda : \begin{cases} \Lambda^* : x = \alpha(r) \\ u = \kappa(r) \end{cases} \\ p = \sigma(r) \ , \end{cases} \\ r = (r_1,\ldots,r_{n-1}) \in J^{n-1} \\ \alpha, \ \kappa \in C^2(J^{n-1}) \ , \ \sigma \in C^1(J^{n-1}) \\ \text{range of } \dfrac{\partial(\alpha)}{\partial(r)} = n-1, \ r \in J^{n-1} \quad , \end{cases}$$

$$(8,18b)=(8,6b) \quad \begin{cases} F(\alpha(r), \ \kappa(r), \ \sigma(r)) = 0 \\ \sum_{h=1}^{n} \sigma_h \ \alpha_{h,i} - \kappa_i = 0, \ i = 1,\ldots,n-1; \ r \in J^{n-1}, \end{cases}$$

(8,18c)=(8,17) $\Delta = 0 \ , \ r \ \varepsilon \ J^{n-1}$,

be a singular initial hyperstrip; and assume

(8,19) $E^{o}(r) \subset \Sigma.$

Then $E^{o}(r)$ is a characteristic hyperstrip.

Note that for the validity of the assertion the first of conditions (18b) is superfluous, as it follows from (19).

Proof. By (19), the pairs (u;p) of the hyperstrip $E^{o}(r)$ are the same as those of Σ,

(8,20) $$u = \phi(\alpha(r)) = \kappa(r), \ p_i = \phi_i(\alpha(r)) =$$

$$= \sigma_i(\alpha(r)), \ i = 1,\dots,n.$$

On the other hand, since $\Delta = 0$ and the rank of the matrix $\frac{\partial(\alpha)}{\partial(r)}$ is (n-1), the last row of Δ is a linear combination of the other (n-1), that is, there exists $\gamma_1(r), \ \dots \ ,\gamma_{n-1}(r)$ such that

(8,21) $$F_j(\alpha(r),\sigma(r)) = \sum_{k=1}^{n-1} \gamma_k(r) \cdot \alpha_{j,k}(r) \ .$$

Combining (20) and (21), we obtain the remaining elements of the last row of Q of (15),

(8,22a) $$\sum_{h=1}^{n}\sigma_h(r) \hat{F}_h = \sum_{h=1}^{n} \sigma_h \sum_{k=1}^{n-1} \gamma_k \alpha_{h,k} =$$

$$\sum_{k=1}^{n-1} \gamma_k \cdot \sum_{h=1}^{n} \frac{\partial\phi}{\partial x_h} \cdot \frac{\partial x_h}{\partial r_k} = \sum_{k=1}^{n-1} \gamma_k \kappa_k \ .$$

$$(8,22b) \quad -F_j - \sigma_j F_u = \frac{\partial F}{\partial x_j} - F_j - \sigma_j F_u = \sum_{h=1}^{n} \hat{F}_h \, p_{h,j} =$$

$$= \sum_{h=1}^{n} \sum_{k=1}^{n-1} \gamma_k \, \alpha_{h,k} \, p_{j,h} = \sum_{k=1}^{n-1} \gamma_k \cdot \sum_{h=1}^{n} \frac{\partial x_h}{\partial r_k} \cdot \frac{\partial p_j}{\partial x_h} =$$

$$= \sum_{k=1}^{n-1} \gamma_k \, \sigma_{j,k} \; .$$

For all these functions it is understood that $r = (r_1, \ldots, r_{n-1})$ is the independent variable and the equalities are valid for all $r \in J^{n-1}$. Thus, $\frac{\partial F}{\partial x_j}$ is the composed partial derivative of $F(x,u,p)$ with respect to the independent variable x_j, and it is assumed also that all of the arguments x,u,p are functions of r. The formulas (20) insure the existence and allow an easy manipulation of the partial derivatives of κ and σ of the hyperstrip in terms of the derivatives of u and p of the integral surface with respect to x. In particular

$$p_{h,j} = \frac{\partial^2 \phi}{\partial x_h \partial x_j} = p_{j,h} \; .$$

Now, (21) and (22) imply that for Q of (15), setting $m = n-1$, the last row is a linear combination of the other $(n-1)$, and therefore, the rank of Q is $n-1$. Since (16) is also satisfied, it follows from assertion 8.1, that the singular initial hyperstrip $E^o(r)$ is characteristic, q.e.d.

From theorem 8.1 and assertions 8.2, 8.3 follows the corollary:

Corollary 8.1 If a hyperstrip is contained in two distinct integral surfaces (with their tangent hyperplanes), then it is characteristic. Infinitely many integral surfaces pass through a characteristic hyperstrip.

Of course, considering a characteristic hyperstrip B, and cutting it by a transverse section, a transverse strip of (n-2)-dimensional support is obtained, through which infinitely many transverse initial hyperstrips can be constructed, each of which determines an integral surface containing B.

The following theorem establishes the most important properties of a singular initial hyperstrip.

Theorem 8.3. Let $E^{o}(r)$, given by (18,a,b,c), be a singular initial hyperstrip.

I. Let n be the rank of M,

$$M = \begin{pmatrix} \alpha_{1,1} & \cdots\cdots & \alpha_{n,1} & \cdots & \kappa_1 \\ \cdot & \cdots & \cdots & \cdots & \cdot \\ \alpha_{1,n-1} & \cdots\cdots & \alpha_{n,n-1} & \cdots & \kappa_{n-1} \\ \hat{F}_1(\alpha(r),\kappa(r),\sigma(r)) & \cdots & \hat{F}_n & & \sum_{h=1}^{n} \sigma_h \hat{F}_h \end{pmatrix}.$$

Then the singularity is evitable.

II. Let n-1 be the rank of M.

A. If the equation $F = 0$ is quasi-linear, then the manifold Λ is a characteristic hyperstrip.

B. _Suppose_ $F = 0$ _is not quasi-linear._

1. _If the rank of_ Q _given by_ (15) _with_ $m = n-1$, _is_ $n-1$, _then_ $E^o(r)$ _is a characteristic hyperstrip and there are infinitely many integral surfaces through it._

2. _If the rank of_ Q _is_ n, _then there is no integral surface containing_ $E^o(r)$.

Proof. I. If the rank of M is n, then in the space $\{0;x,u\}$ the initial hyperstrip $E^o(r)$ is actually transverse, and hence, by a change of coordinates (x,u), we can obtain a regular Cauchy problem. In this sense, the singularity is evitable. However, if we deal with a strip $E^o(r)$, and not a curve Λ, we must generalize in some sense the definition of strip, since p becomes infinite (see exercise 9 of ∮ 3).

II. A.The proof is the same as the 2-dimensional case and is left as Exercise 4.

B. 1. This is an immediate consequence of the preceeding assertions and their corollary.

B. 2. If there were an integral surface, by assertion 8.3, $E^o(r)$ would be characteristic, and the rank of Q would be $n-1$. Therefore, there is no such integral surface containing $E^o(r)$, q.e.d.

This last case is particularly interesting. Through each element of $E^o(r)$, we can construct the characteristic strip, thus obtaining a surface Φ. Why is Φ not an integral surface through $E^o(r)$? To better understand this situation, we examine the proofs of the existence of a solution in Theorem 8.2 and of Assertion 8.3. In the proof of Theorem 8.2.II, the transversality condition is used to prove the existence of the tangent hyperplane, since even if the functions ξ, ζ, π are of class one, if the directions $\alpha_{,i}$ are not transverse to $\frac{dx}{ds}$, they do not define a tangent hyperplane. In the same theorem it is also used to put the parametric equation in explicit form $u = \phi(x)$.

The proof of Theorem 8,3 would wholely apply to the presently considered hyperstrip $E^{o}(r)$, which satisfies all the required conditions, and would apply to Φ instead of Σ, if only Φ were in explicit form and of class two.

Therefore, we forsee that in a neighborhood of a point of Λ in general it will not be possible to put Φ in explicit form, but should it be possible in this neighborhood, it would not be of class two.

Theorem 8.3 does not exclude in the last case the following situation which will occur, in general, at each point $(\tilde{x},\tilde{u}) \in \Lambda$.

Suppose in the solution of the Cauchy problem given by (8) we limit variation of s by

$$0 < s - s_o < \delta_2(r).$$

Thus, we obtain a surface $\Psi : u = \psi(x)$, defined in an open G. Ψ is an integral surface of $F = 0$, originating at $E^{o}(r)$, although not containing this singular initial hyperstrip. But there is a neighborhood of $\tilde{x} \in \overset{*}{\Lambda}$, call it $\tilde{\Lambda}$, such that $\tilde{\Lambda}$ lies in ∂G. We now construct a closed C with nonempty interior with smooth boundary ∂C such that $C \subset \tilde{\Lambda} \cup G$ and that there is a neighborhood of $\tilde{x}$ in ∂C. We may continue Ψ and define it on C in such a way that it exists for $s = s_o$ and contains the initial hyperstrip $E^o(r)$ in a neighborhood of $(\tilde{x},\tilde{u})$. We may also extend the definition of integral surfaces to closed domains with sufficiently smooth boundaries. Then $u = \psi(x)$, $x \in C$, is explicit, continuous, may have a tangent hyperplane at all points and continuous first derivatives , and is an integral surface at each interior point of C. But, it cannot have continuous second derivatives on ∂C in a neighborhood of $(\tilde{x},\tilde{u})$, since if it did, by Assertion 8.3, $E^o(r)$ would be a characteristic hyperstrip contrary to hypothesis (see Exercise 11 of § 5.).

Exercises

1. Carry out the proof of Theorem 8.1.

2. Prove that the strip conditions are invariant with respect to a change of coordinates.

[Let (10) be the strip whose equations satisfy the strip condition (6b) with $i = 1,2, \ldots ,m$. Make the change $r = \rho(t_1, \ldots ,t_m)$ and verify that the new equations satisfy the new strip conditions.]

3. Suppose we are given 2n-m first integrals $w_1, \ldots ,w_{2n-m}$ of the characteristic system, $1 \leqq m \leqq n$, which are functionally independent, and also independent of F. Eliminate $p_1, \ldots ,p_n$ from the system

$$F = 0,\ w_i(x,u,p) = C_i\ ,\ i = 1, \ldots ,2n - m\ .$$

Prove that we obtain a system of characteristic strips with m-dimensional support depending on (2n-m) aribitrary constants.

4. Carry out the proof of Part II,A. of Theorem 8.3.

§ 9. Hamilton-Jacobi Method.

1. Characteristic System and Complete Integrals.- Consider an equation with (n+1) independent variables $(x_1, \ldots, x_n, s) = (x, s)$ in Hamiltonian form

$$(9,1) \qquad u_s + H(x,s,p) = 0 \ ,$$

$$u_s = \frac{\partial u}{\partial s}, \ p = (p_1, \ldots, p_n), \ p_i = \frac{\partial u}{\partial x_i} \ ,$$

$$i = 1, \ \ldots \ , n \ ,$$

where u is the unknown function. The Hamiltonian form is of course a particular case of the general form F = 0. Therefore, we assume for the first member as well as for the integral surfaces $u = \phi(x,s)$, the regularity conditions of article 1, § 8.

The function H is called the Hamiltonian. In equation (1), the variable s is distinguished, in the sense that the equation is solved for u_s; moreover, the equation does not depend directly on u, but only on its derivatives.

That a first order partial differential equation is in Hamiltonian form does not imply a loss of generality. Any first order equation can easily be put in this form through a simple transformation, but at the cost of adding another independent variable.

Actually, consider the equation with n-independent variables in the unknown v

$$F(x,v,q) = 0,\ F_v \neq 0,$$

$$(9,2) \qquad x \equiv (x_1,\ldots,x_n),\ q \equiv (q_1,\ldots,q_n) \equiv \left(\frac{\partial v}{\partial x_1}, \ldots , \frac{\partial v}{\partial x_n}\right).$$

If $f_v = 0$, it suffices to solve for one of the derivatives q_i and the equation is then in Hamiltonian form.

We apply the Jacobi-transformation, thus generalizing Article 1, § 2. We assume v is implicitly defined by a first integral, independent of p, of the characteristic system. Call it $u(x,v)$ and set

$$(9,3a) \qquad u(x,v) = c,\ \frac{\partial u}{\partial v} \neq 0,\ u \in C^2(D^{n+1}) ,$$

which implicitly defines a 1-parameter system of integral surfaces $v = \phi(x,c)$.

Taking derivatives of (3a) with respect to x_i, and letting u_i be the partial derivative of u with respect to x_i, we have

$$u_i + u_v\, q_i = 0 ,\ i = 1, \ldots ,n .$$

Therefore, equation (2) may be written

$$(9,3b) \qquad F\left(x,v, \frac{-u_1}{u_v} , \ldots , \frac{-u_n}{u_v}\right) = G(x,v,u_1,\ldots,u_n,u_v) = 0 ,$$

Thus we have obtained an equation in u depending on (n+1) independent variables. Solving for one of the derivatives, we obtain the Hamiltonian form; if $\frac{\partial G}{\partial u_v} \neq 0$ (see Exercise 1.), it suffices to solve for u_v and set $v = s$ to obtain (1).

The characteristic system takes a simpler form with notation analogous to that established in § 8, article 1. The characteristic system of (1) is

$$(9,4) \qquad \frac{dx_1}{\hat{H}_1} = \dots = \frac{dx_n}{\hat{H}_n} = \frac{ds}{1} = \frac{du}{p_1\hat{H}_1+\dots+p_n\hat{H}_n-H} =$$

$$= \frac{dp_1}{-H_1} = \dots = \frac{dp_n}{-H_n} = \frac{du_s}{-H_s} .$$

This system of (2n+2) equations in the (2n+2) unknown x,u,p,u_s, which are functions of s, can be decomposed into the regular, normal system

$$(9,5) \qquad \frac{dx_i}{ds} = \hat{H}_i(x,s,p), \quad \frac{dp_i}{ds} = H_i(x,s,p), \quad i = 1,,\dots,n$$

and the two equations

$$(9,6) \qquad \frac{du}{ds} = \sum_{k=1}^{n} p_k \hat{H}_k - H , \quad \frac{d\,u_s}{ds} = -H_s ,$$

The first reduces to a quadrature after (5) has been integrated, while the second reduces to $u_s + H =$ constant.

The ordinary differential systems, such as (5), which derive from a unique function or Hamiltonian in the same way that (5) derives from a function $H(x,s,p)$, are called <u>canonical differential systems</u>.

The canonical system (5) has a general integral

$$(9,7) \qquad x = \xi(s,c), \quad p = \pi(s,c), \quad c = (c_1,\dots,c_{2n})$$

depending on 2n constants. With the other 2 integration constants from (6), we obtain the system of strip solutions of (4). Among these, the integral ones constitute the characteristic strips; they are those satisfying $u_s + H = 0$.

A complete integral of the equation must contain as many arbitrary parameters as independent variables. Hence a complete integral of (1) depends on (n+1) parameters. Equation (1) does not depend directly on u. Therefore every complete integral can be put in a form containing an additive constant, changing parameters if necessary (see Exercise 2). We may therefore adopt the following definition with no loss of generality.

A <u>complete</u> <u>integral</u> of an equation in Hamiltonian form is a system of integral surfaces

$$(9,8a) \qquad u = \phi(x,s,a) + a_o \ , \ a = (a_1,\dots,a_n),$$

$$(x,s) \ \varepsilon \ G, \ (a,a_o) \ \varepsilon \ K, \ \phi \ \varepsilon \ C^2(GxK),$$

depending on (n+1) independent parameters.

For (8a) to have (n+1) independent parameters, it is necessary and sufficient that the matrix

$$\frac{\partial(\phi + a_o, \ \phi_1,\dots,\phi_n,\phi_s)}{\partial(a,a_o)}$$

has rank (n+1)(see exercise 3). This condition is evidently equivalent to

$$\text{rank} \quad \frac{\partial(\phi_1, \ \dots,\phi_n, \ \phi_s)}{\partial(a)} \quad = n.$$

But from the form of equation (1), if follows (Excerise 4) that in this horizontal matrix the last column is a combination of the others, and hence, follows the next assertion.

Assertion 9.1 The system of integral surfaces of (8a) is a complete integral of (1) if and only if

$$(9,8b) \qquad \left| \frac{\partial(\phi_1, \ldots , \phi_n)}{\partial(a)} \right| \neq 0 \; .$$

It is easily verified (Exercise 5) that conversely, the complete integral (8a,b) yields by elimination of the (n+1) constants (a, a_o) a first order differential equation in Hamiltonian form. For simplicity sake, we call the function $\phi(x,s,a)$ alone a complete integral.

2. The Hamilton-Jacobi Theorem.- As in the 2-variable case, we summarize the most important properties of complete integrals in the following 2 theorems.

Theorem 9.1. Let $\phi(x,s,a)$, given by (8a,b), be a complete integral of (1). Then we have:

I. 1) The system of functions

$$(9,9) = (9,7) \begin{cases} x = \xi(s,a,b) \\ p = \pi(s,a,b), \\ a = (a_1,\ldots,a_n) \;,\; b = (b_1,\ldots,b_n), \end{cases}$$

depending on 2n arbitrary parameters, implicitly defined by the relations

$$(9,10)\quad \begin{cases} \dfrac{\partial \phi}{\partial a_i} = b_i \\[2ex] p_i = \dfrac{\partial \phi}{\partial x_i} \equiv \phi_i \ , \ i = 1,\ldots,n, \end{cases}$$

constitutes the general integral of the canonical differential system (5).

2) *The system of functions constituted by* (9) *and*

$$(9,11)\quad \begin{cases} u = u_o + \int_{s_o}^{s} (\pi_1 \hat{H}_1 + \ldots + \pi_n \hat{H}_n - H)\, ds \\[2ex] u_s = -H(\xi(s,a,b),s,\pi(s,a,b)), \end{cases}$$

depending on (2n+1) *arbitrary constants constitutes the set of all characteristic strips of equation* (1).

II. *Given an arbitrary relation among the* (n+1) *parameters,* $a_o = \mu(a)$, *the surface* $u = \psi(x,s)$ *obtained through elimination of* a *in*

$$(9,12)\quad \begin{cases} u = \phi(x,s,a) + \mu(a) \\[1ex] \dfrac{\partial \phi}{\partial a_i} + \mu_i = 0 \ , \ i = 1,\ldots,n, \end{cases}$$

is an integral surface of (1), *provided there is a real point* (x^o,s_o) *for* $a = a^o$ *and the Jacobian* $\dfrac{\partial(\phi_a)}{\partial a} \neq 0$ *at* (x^o,s_o,a^o).

Proof. I. To pass from equations (10), which are given, to (9), we may proceed thus. Solve the first group of n equations of (10) for the x_i as functions of s,a,b obtaining

$$x = \xi(s,a,b),$$

which is the first of (9). The solving for the x_i is certainly possible since

$$\frac{\partial\left(\frac{\partial\phi}{\partial a_1}, \dots, \frac{\partial\phi}{\partial a_n}\right)}{\partial(x)} \neq 0,$$

which is the same as (8b). Having obtained the x_i, substitute them in the second of (10) obtaining $p_i = \phi_i(\xi(s,a,b),s,a)$, that is to say, $p = \pi(s,a,b)$, which is the second of (9).

To prove that the functions obtained are solutions of (5), we first consider the first group of (10) and (1), from which it follows, taking derivatives, that

$$\frac{\partial}{\partial s}\left(\frac{\partial\phi}{\partial a_i}\right) - \frac{\partial}{\partial a_i}(u_s + H(x,s,p)) =$$

$$(9,13) \qquad \frac{\partial^2\phi}{\partial a_i \partial s} + \sum_{k=1}^{n} \frac{\partial^2\phi}{\partial a_i \partial x_k}\frac{\partial\xi_k}{\partial s} - \left(\frac{\partial^2\phi}{\partial s \partial a_i} + \sum_{k=1}^{n} \frac{\partial H}{\partial p_k}\frac{\partial}{\partial a_i}\left(\frac{\partial\phi}{\partial x_k}\right)\right)$$

$$= \sum_{k=1}^{n} \frac{\partial^2\phi}{\partial a_i \partial x_k}\left(\frac{\partial\xi_k}{\partial s} - \hat{H}_k\right)$$

$$= 0, \quad i = 1, \dots, n.$$

In this deduction, the b_i are considered constant, and x,s,a of the complete integral are taken as independent variables; but, to prove the functions ξ satisfy (5), we must observe that the x's are functions of s in such a way that the first partial derivative of x with respect to s must be understood as a composed partial derivative; moreover, we have applied the second group of formulas of (10). Now the last equality of (13) may be considered as a homogeneous linear algebraic system of n equations whose matrix of coefficients has non-zero determinant. Therefore, the $x = \xi(s,a,b)$ satisfy the first group of equations of (5).

To prove $p = \pi(s,a,b)$ satisfies the second group of (5), we have, in the same way, taking derivatives of (10) and (1), that

$$\frac{\partial \pi_i(s,a,b)}{\partial s} = \frac{\partial}{\partial s}\left(\frac{\partial \phi}{\partial x_i}\right) - \frac{\partial}{\partial x_i}(u_s + H(x,s,p))$$

$$= \frac{\partial^2 \phi}{\partial x_i \partial s} + \sum_{k=1}^{n} \frac{\partial^2 \phi}{\partial x_i \partial x_k} \frac{\partial \xi_k}{\partial s} -$$

$$\left(\frac{\partial^2 \phi}{\partial s\, \partial x_i} + H_i + \sum_{k=1}^{n} \frac{\partial}{\partial x_i}\left(\frac{\partial \phi}{\partial x_k}\right)\right),$$

$$i = 1, \ldots, n.$$

Considering we have already proven the theorem for the first group, the preceeding expression can be simplified. What remains expresses precisely that $\pi(s,a,b)$ satisfies the second group of (5).

II. The second part is a consequence of the first. Actually, if in (12) we give a a fixed value $a = a^*$, and set $\mu_i(a^*) = -b_i^*$, then the second of (12) becomes the first of (10), that is to say,

$$(9,14a) = (9,9) \qquad x = \xi(s,a^*,b^*).$$

Taking derivatives of the first of (12) gives

$$(9,14b) = (9,9) \qquad p = \pi(s,a^*,b^*),$$

in such a way that for $a = a^*$, (12) represents a characteristic curve that together with (14b) and (11) constitutes a characteristic strip. Hence, if we solve the second of (12) for

a_i, which is always possible since the Jacobian does not vanish, and substitute in the first, we obtain a real surface in explicit form generated by an n-parameter system of characteristic strips. Therefore, there is, in fact, an integral surface, q.e.d.

In (12), we assumed for simplicity sake a relation of the form $a_o = \mu(a)$. It is easily verified that the proof remains the same if instead we assume $a_n = \mu(a_1, \ldots, a_{n-1})$, or even $a_n = \mu(a_o, \ldots, a_{n-1})$. In this case, the partial derivatives with respect to a parameter must be understood as composed partial derivatives.

Note, we could have applied Corollary 8.1 and proven this second part in entirely the same manner as the 2-variable case, and then, deduced I from II. Here, we preferred a direct proof of I, which constitutes the theorem of Hamilton-Jacobi.

Their idea was to invert the process followed in this chapter. We have reduced the problem of the integration of first order partial differential equations to the solution of a system of ordinary differential equations, which we assume has been previously solved.

By the introduction of the Hamiltonian, the reduction is done not to arbitrary systems, but precisely to canonical differential systems. The contribution of Hamilton-Jacobi consists in having established that conversely, the integration of a canonical differential system may be reduced to knowledge of a complete integral of a first order partial differential equation in Hamiltonian form. Now, in the study of dynamical systems, one often arrives, through the Lagrange equations as well as variational principles, at canonical differential systems in which it is easy to obtain a complete integral of the equation in Hamiltonian form, equivalent to the canonical system. Then the theorem just proven supplies a systematic method for the resolution of the corresponding dynamics problem. For an example see Ex. 10.

3. <u>The Cauchy Problem</u> - Given a complete integral, the solution of the Cauchy problem offers no difficulty.

Suppose we are given a sufficiently smooth transverse initial hyperstrip $E^o(r)$, which, since there are $(n+1)$ independent variables, must have n-dimensional support. Call it

$$(9,15a) \quad E^o(r)\begin{cases} \Lambda \begin{cases} \Lambda^* : x = \alpha(r),\ s = \alpha_{n+1}(r),\ r = (r_1,\ldots,r_n) \\ u = \kappa(r) \end{cases} \\ p = \sigma(r),\ u_s = \sigma_{n+1}(r),\ r \in J^n , \end{cases}$$

which is assumed to satisfy (1) and the strip conditions

$$(9,15b) \begin{cases} \sigma_{n+1}(r) + H(\alpha(r), \alpha_{n+1}(r), \sigma(r)) = 0 \\ \sum_{k=1}^{n+1} \sigma_k \alpha_{k,i} - \kappa_i = 0,\ i = 1,\ldots,n ,\ r \in J^n , \end{cases}$$

and the transversality condition

$$(9,15c) \quad \Delta = \begin{vmatrix} \alpha_{1,1} & \cdots & \alpha_{n,1} & \alpha_{n+1,1} \\ \cdot & \cdots & \cdot & \cdot \\ \alpha_{1,n} & \cdots & \alpha_{n,n} & \alpha_{n+1,n} \\ \hat{H}_1(\alpha(r),\alpha_{n+1}(r),\sigma(r)) & \cdots & \hat{H}_n & 1 \end{vmatrix} \neq 0, r\varepsilon J^n .$$

The Cauchy problem is thus posed: to find an integral surface of (1), $u = \psi(x,s)$, (x,s) varying over a domain G of $\{0;x,s\}$ containing Λ^*, and passing through $E^o(r)$ given by (15a,b,c). Assuming a complete integral (8a,b) of (1) is known, the following theorem offers a solution.

Theorem 9.2. Given the transverse hyperstrip $E^o(r)$ of (15a,b,c), we obtain the integral surface $u = \psi(x,s)$, $(x,s) \in G$, through $E^o(r)$ applying the second part of Theorem 9.1. As a relation binding a, a_o, we must take that which results from eliminating r in

$$(9,16) \quad \begin{cases} \kappa(r) = \phi(\alpha(r), \alpha_{n+1}(r), a) + a_o \\ \sigma_i(r) = \phi_i(\alpha(r), \alpha_{n+1}(r), a), \; i = 1,\ldots,n \\ \sigma_{n+1}(r) = \phi_s(\alpha(r), \alpha_{n+1}(r), a). \end{cases}$$

Proof. By Theorem 8.2, there exists a unique integral surface solution of our problem, and moreover, it is obtained by constructing the characteristic strip through each element of $E^o(r)$. Therefore, by I of Theorem 9.1, there is certainly a relation between a, a_o, which by II, must yield the surface $u = \psi(x,s)$.

The relation between the (n+1) parameters $a_1,\ldots,a_n,a_o$ must make the equations $u = \phi(x,s,a) + a_o$, the second group of (10), and $u_s = \phi_s(x,s,a)$, compatible with the very same equations for particular values at the points of $E^o(r)$, that is to say, with (15a). But (16) establishes precisely this compatibility, and eliminating the n-parameters $r_1,\ldots,r_n$ from the (n+2) equations, we obtain the desired relation between $a_1,\dots,a_n,a_o$, q.e.d.

We could give a direct proof without appeal to Theorem 8.2. To that it is convenient to make the change of variables $r \to x$, that is to say, take $x_1,\dots,x_n$ as independent variables instead of $r_1,\dots,r_n$; this is always possible since the hyperstrip with $x_i = r_i$, $s = \alpha_{n+1}(r)$ is always evidently transverse (Exercise 7). In (16), we can prescind the last equation. The n equations

$$\sigma_i(x) = \Phi_i(x, s(x), a)$$

can certainly be solved for $a_1,\dots,a_n$ due to (8b), and we obtain

$$a_k = \tau_k(x) \ , \ k = 1, \dots , n.$$

Now if the Jacobian $\frac{\partial \tau}{\partial x}$ is non-zero, we can solve these equations for the $x_i = \theta_i(a)$, which substituted in the first of (16) yields the desired relation of the form $a_o = \mu(a)$. If on the contrary $\frac{\partial \tau}{\partial x} = 0$, then there is a relation between the second members $\mu(\tau(x)) = 0$, and the desired relation $\mu(a) = 0$ is independent of a_o.

Exercises.

1. Prove that G of (3b) is independent of u_v if and only if F is homogeneous with respect to p.
2. Prove that any complete integral of (1) depending on (n+1) independent parameters can, by a change of parameters, be put in a form having a parameter in additive form. [See Exercise 1. of ϕ 6.].
3. Prove the necessity and sufficiency of the condition

stated in 1 that $\phi(x,s,a) + a_o$ has $(n+1)$ independent parameters.

4. Prove that in the horizontal matrix

$$\frac{\partial(\phi_1, \ldots, \phi_n, \phi_s)}{\partial(a)}$$

the last column is a linear combination of the others.

5. Verify that the elimination of the constants in a complete integral (8a,b) leads to an equation in Hamiltonian form (1).

6. Generalize Exercise 2 of § 6 to the n-variable case.

7. Let the base initial curve Λ^* in (15a) be given by

$$x_i = r_i \ , \ i = 1,\ldots,n \ , \ s = \alpha_{n+1}(r) \ , \ \alpha_{n+1} \in C^2(J^n).$$

Prove that in (15c),

$$\Delta = (1-n) \neq 0 \ .$$

8.a. Given an equation in Jacobi form, that is,

$$J(x,t,w,w_1,\ldots,w_n,w_t) = 0, \ x = (x_1,\ldots,x_n), \ w_i = \frac{\partial w}{\partial x_i} \ ,$$

where $J_w = 0$ and J is a homogeneous function of degree m in $w_1,\ldots,w_n,w_t$, show that it can be reduced to another of general type with one less independent variable,

$$F(x,u,p) = 0, \ p = (p_1,\ldots,p_n), \ p_i = \frac{\partial u}{\partial x_i} \ ;$$

and conversely.

[Set $t = u$, $w(x,u) = c \equiv$ const, $w_t \, p_i + w_i = 0$; $F(x,u,p) = J(x,u,w,-p_1,\ldots,-p_n,1)$; $J(x,t,w,w_1,\ldots,w_n,w_t) =$ $= w_t^{\,m} . \ F(x,t, -\frac{w_1}{w_t} \ , \ \ldots \ , -\frac{w_n}{w_t} \ )$. Note moreover that the Jacobi form is more restrictive than the Hamiltonian form.]

b. Prove also that if $w = w(x,t)$ is a solution of $J = 0$, the 1-parameter system of surfaces $u = \varphi(x,c)$, implicitly defined by $w(x,u) = c = \text{constant}$, is a 1-parameter system of integral surfaces of $F = 0$. Conversely, if $u = \varphi(x,c)$ is a 1-parameter system of integral surfaces of $F = 0$, then solving for c and obtaining $c = w(x,u)$, it follows that $w = w(x,t)$ is an integral surface of $J = 0$.

[See § 2]

c. Finally, let $E^o = (x^o, t_o, p^o)$ be an integral element of $F = 0$, and $\overline{B}(s;s_o,\overline{E}^o)$ be the characteristic strip of $J = 0$ satisfying $\overline{B}(s_o;s_o,\overline{E}^o) = \overline{E}^o = (x^o, t_o, w_o, -\lambda p_1^o, \ldots, -\lambda p_n^o, \lambda)$.

Prove that suppressing in the strip $\overline{B}$ the equations $w = \zeta(s;s_o,\overline{E}^o)$ and $w_t = \theta(s;s_o,\overline{E}^o)$ and setting $\lambda = 1$, we obtain the characteristic strip $B(s;s^o,E^o)$ of $F = 0$ satisfying $B(s_o;s_o,E^o) = E^o$.

[Remark, in the strip $\overline{B}$, $\frac{\partial x_i}{\partial w_o} = \frac{\partial w_i}{\partial w_o} = 0$, $\frac{\partial w}{\partial w_o} = 1$.

Applying the formulas for solutions of (a) and the homogeneity relation, put the characteristic system of $J = 0$ in terms of that of $F = 0$ and verify that the statement is fulfilled.]

9. Suppose we are given the equation $F(x,u,p) = 0$ and $u = \phi(x)$ is an integral surface, and suppose we know either the general integral of the characteristic system or a complete integral. Determine a 1-parameter system of integral surfaces

$$u = \psi(x,a)$$

such that

$$\psi(x,0) = \varphi(x)$$

[If we solve for a in $u = \psi(x,a)$ and obtain $a = w(x,u)$, then $w(x,u)$ is a first integral of the characteristic system of $J = 0$, that is, $F = 0$ in Jacobi form. The problem may be solved either by elimination processes or by taking a transverse initial strip in the given surface, for example $u = \varphi(x) = 0$, $p_i = \varphi_i(x,0)$, and preserving the same Λ^*, $\varphi(x) = 0$ and modifying the data u through summands or factors depending on arbitrary parameters and determined functions.]

10. As an example of an application of the Hamilton-Jacobi method to the solution of dynamical systems, we consider the spherical pendulum, that is to say, the motion of a material point of mass m, constrained to move without friction under the field of gravity in a sphere of radius R. We take as origin of a cartesian coordinate system $\{O;x,y,z\}$ the center of the sphere and as direction and positive sense of the z-axis, those of gravity, Figure 3. We assume the trajectory originates at $t = 0$ from an initial point P with an initial velocity v_o, assumed to have the direction and positive sense of the tangent at P to the parallel through P, which does not imply any loss of generality, for whatever trajectory results, the z coordinate always has at least one maximum or minimum. We take the x-axis in such a way that the plane xOz passes through P and introduce the spherical coordinates

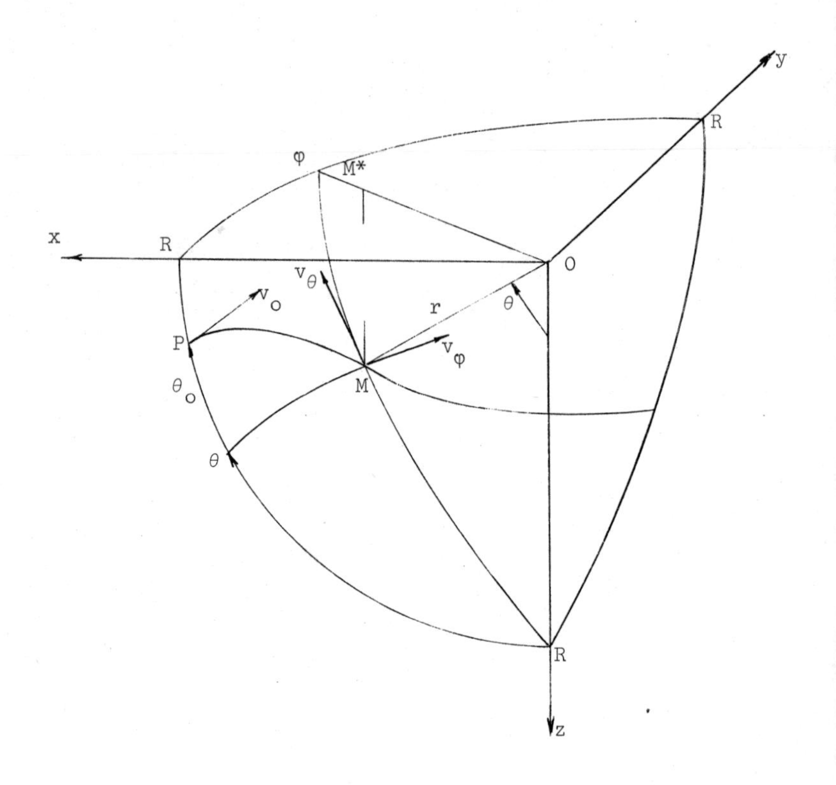

Figure 3.

$$(x,y,z) = (r,\theta,\varphi)$$

and take θ,φ as generalized coordinates of the motion. Of course, all the coordinates are functions of the independent variable t, time. Therefore, the initial parameters are

$$\theta(0) = \theta_o, \ \varphi(0) = 0, \ v_\theta'(0) \equiv \frac{dv_\theta(0)}{dt} = 0, \ v_\varphi'(0) = v_o \ .$$

The kinetic energy T and potential energy V are easily obtained and given by

$$T = \frac{1}{2} m(v_\theta^2 + v_\varphi^2) = \frac{1}{2} m(R^2\theta'^2 + R^2 \sin^2\theta \cdot \varphi'^2)$$

$$V = -mgz = - mg R \cos\theta \ .$$

By the Lagrange equations

$$\frac{d}{dt} \frac{\partial L}{\partial q_i'} = \frac{\partial L}{\partial q_i}, \ q_1 = \theta, \ q_2 = \varphi \ ,$$

$$L(q_1,q_2,q_1',q_2') = T(q_1,q_2,q_1',q_2') - V(q_1,q_2)$$

we obtain the equations of motion

$$R_\theta'' = R \sin\theta \cos\theta \cdot (\varphi')^2 - g \sin\theta$$

$$(R^2 \sin^2\theta \cdot \varphi')' = 0.$$

The second yields immediately the first integral

$$R^2 \sin^2\theta \cdot \varphi' = R \sin\theta_o \cdot v_o \equiv k;$$

and from it, multiplying the first equation by $\theta' d\theta$ and integrating from P, we obtain another first integral

$$R^2 \theta'^2 + \frac{k^2}{R^2 \sin^2\theta} - 2g R \cos\theta = v_o^2 - 2Rg \cos\theta_o \equiv$$

$$v_o^2 - 2gz_o \equiv c.$$

Finally, this equation can be integrated in finite terms by means of elliptic functions.

Let us see how the Hamilton-Jacobi method yields the same result. It is perhaps a bit more complicated, but solves the problem more systematically and offers new connections with other theories. We introduce the _generalized momenta_ (also called _canonical_ or _conjugate_ momenta) p_1, p_2 by

$$p_1 = \frac{\partial L}{q_1'} = m R^2 \theta', \quad p_2 = \frac{\partial L}{q_2'} = mR^2 \sin^2\theta \cdot \varphi' .$$

The _Hamiltonian_ H in the generalized coordinates and momenta takes the form

$$H(\theta,\varphi;p_1,p_2) = T + V = \frac{p_1^2}{2mR^2} + \frac{p_2^2}{2mR^2\sin^2\theta} - gmR\cos\theta.$$

We call the function $u = u(\theta,\varphi,t)$ such that $u_t = H$, the _generating function_. (H represents the total energy of the point and remains constant during the motion). We obtain the _Hamilton-Jacobi equation_ with three independent variables

$$(*) \qquad u_t + H(\theta,\varphi;p_1,p_2) = 0, \quad p_1 = \frac{\partial u}{\partial\theta}, \; p_2 = \frac{\partial u}{\partial\varphi} .$$

Verify that the canonical differential system (5) corresponding to this equation is equivalent to the equations of motion obtained starting from the Lagrangian L.

In the case of equation (*), we can obtain systematically a <u>principal function</u> of Hamilton, or complete integral of (*), say $u = S(\theta,\varphi;a_1,a_2) + a_o$, because H depends explicitly on the only variable θ. It follows that the variables can be separated, that is, we can obtain, only through quadratures, one complete integral S of the form

$$S = S_1(\theta;a_1,a_2,a_o) + S_2(\varphi;a_1,a_2,a_o) + S_3(t;a_1,a_2,a_o).$$

Find $W(\theta)$ such that

$$u = W(\theta;c,k,a_o) + k\varphi - ct$$

is a complete integral of (*). With the obtaining of this complete integral one has substantially solved the problem of integration of the equation of motion of the spherical pendulum thanks to Theorem 9.1. Observe that the possibility of separating variables depends on the independent variables which have been chosen or on a convenient change of canonical variables.

Prove that the areolar velocity of the projection of the rod of the pendulum on the plane $\{0;x,y\}$ is constant.

Is it possible that the pendulum bob describes a parallel of the sphere? $[v_o^2 = g \cdot R \sin\theta_o \cdot \mathrm{tg}\,\theta_o]$.

Prove that the reaction the moving point exerts on the

sphere is the sum of the centrifugal force and of the component of the weight of the point on the normal to the sphere.

Bibliographical Note

Courant, R. and Hilbert, D.: Methods of Mathematical Physics; Volume II: *Partial Differential Equations*, Interscience, New York, 1962.

Goursat, E.: *Cours d'Analyse Mathématique*, Volume II, Gauthier-Villars, Paris, 1915. Many editions; also in English (Dover, New York, 1959).

ANALYTICAL INDEX